POWER CONTROL

CONTROL

WITH

SOLID-STATE

DEVICES

POWER CONTROL
WITH
SOLID-STATE DEVICES

IRVING M. GOTTLIEB

 TAB Professional and Reference Books

Division of TAB BOOKS Inc.
P.O. Box 40, Blue Ridge Summit, PA 17214

TAB BOOKS Inc. offers software for sale. For information and a catalog, please contact TAB Software Department, Blue Ridge Summit, PA 17294-0850.

FIRST EDITION
FIRST PRINTING

Library of Congress Cataloging in Publication Data

Gottlieb, Irving M.
Power control with solid-state devices.

Originally published: Reston, Va. : Reston Pub Co.,
©1985.
Includes index.
1. Power electronics. 2. Solid state electronics.
I. Title.

TK7881.15.G68   1987   621.31′7   86-29987
ISBN 0-8306-0795-1

Contents

3 Some Practical Aspects of Solid-State Devices   106

4 Circuit Applications Using Power Transistors and Power ICs   175

5 Applications Using Thyristors   245

6 From the Classic to the Avant-Garde: A Look at Newer Developments   313

Preface

The technological marvel of the mid-twentieth century was the semi-conductor transistor; its simulation of electron-tube amplifying ability was scientifically intriguing because this behavior was accomplished without vacuum and without a hot filament or a heater. Yet the alleged experts extrapolated a rather low-key future for this affront to entrenched tube technology. True, none could deny that the intruder existed both in principle and in hardware. It was begrudgingly and condescendingly acknowledged that the cute little device was probably here to stay, but the consensus seemed to be that its role would be a minor, if not dismal, one. Its use was foreseen in toys, hobby projects, novel flea powered control circuits, and perhaps audio amplification if the objective were something less than high-fidelity reproduction. At best it was to be little more than an adjunct to the tube. Such predictions were also reinforced by contentions that the transistor was fragile, unreliable, and nonlinear and could not be mass-produced with predictable parameters; i.e., it was not worthy of consideration as an engineering device.

It would be easy to say simply that these were the mouthings of false prophets. Yet many of these unenthusiastic prognostications were made by high achievers in science and engineering. It would be more accurate, as well as more equitable, to concede that even experts can go astray when peering into the future. And, we must admit, they had little to go on—early transistors delivered performance constrained by a few tens of milliamperes and volts. Thus power outputs were initially limited to the several-hundred milliwatt region. Frequency capability beyond the audio range was not readily realized. Surely we cannot fault those of the era who were not able to visualize solid-state devices processing kilovolts and kiloamperes, controlling hundreds and thousands of watts of load power, and operating with respectable power-handling capability at tens, and even hundreds, of megahertz. What has transpired since the first few years following the advent of the transistor is nothing less than a technological miracle.

By the early 1970s true power devices were available and a good measure of maturity had set in; many devices and techniques then developed retain their basic validity in design practice even though better devices have been developed. Interestingly one often sees combinations of the older and newer devices and circuitries. For this reason this book deals with both: the widely accepted power control systems and the avant-garde—those devices and applications that emerged during the past several years or so.

It is the author's intent that practical benefits from this book will accrue for all involved in solid-state control and processing of electrical power. Scientists, engineers, and technicians should find valuable insights and guidance in their pursuits and the same should be the case for servicemen, radio amateurs, hobbyists, and experimenters. It is hoped, too, that those in marketing, administration, and quality control will find this book useful for sharpening their technical grasp of the systems and equipments with which they are involved.

The root source of information processed in a book of this kind is derived from the competently staffed and well-equipped research and development laboratories of the semiconductor firms. The assistance rendered by the following individuals and their respective firms is gratefully acknowledged by the author: W. C. Caldwell, distributor sales administrator, Delco Electronics; Neil Cleer, manager, marketing services, International Rectifier Corp.; Walter B. Dennen, manager, news & information, RCA Solid-State Division; Robert Dobkin, director, advanced circuit development, National Semiconductor Corp.; Forest B. Golden, P. E., customer engineering, General Electric Semiconductor Products Department; Dave Hoffman, applications engineer, Siliconix, Inc.; and Lothar Stern, manager, Technical Information Center, Motorola Semiconductor Products, Inc.

IRVING M. GOTTLIEB

Introduction

The logic of the editorial format of this book is one of evolution; from the first chapter through the sixth, the path of the reader commences with basic principles and progresses through increasingly complex applications. This same path progresses up the time scale from widely accepted implementations that had their beginnings in the mid-1960s to newer device and circuit technology emerging on the scene during the late-1970s and early 1980s. Another form of progression has to do with the amount of detail accorded to circuit explanation. As one goes through succeeding chapters, it will be observed that redundant discussions of circuit operation are gradually diminished. It is felt that these progressions enhance the value of the book for instructor and student, for the electronics practitioner just moving into the domain of solid-state power, and for the general-interest reader.

At the same time the various circuitries have a large measure of independence from all other circuit applications. Thus the already knowledgeable reader can locate a desired project in the index and deal with the relevant portion of the text alone.

1

Basic Principles of Solid-State Control of Power

POWER ELECTRONICS

From its inception the wonder of electronics was derived largely from the detection and control of feeble power levels. Admittedly this association stems from early days of wireless communication and from the then-awesome ability of electron tubes to deliver useful outputs from minute antenna currents. Even though giant transmitting tubes and industrial tubes of multikilowatt capability ultimately evolved, a general understanding prevailed concerning the differentiation that one was obliged to make when dealing with radio reception and when dealing with so-called power engineering. It was more or less directly from *radio* that the more comprehensive term *electronics* came into being. Intentionally or otherwise, electronics continued to remain suggestive of the processing of tiny signal levels.

With the further progress in semiconductor technology, devices and applications evolved that readily delivered power greatly exceeding that needed merely to actuate a meter, display, or small transducer. Moreover, these new devices could, by themselves or in conjunction with solid-state circuitry, produce hefty outputs when actuated by low-power signals. What emerged was a new state of the art not accurately described by the phrase *power engineering*. Rather aspects of both classifications clearly applied. Thus, we have the term *power electronics*.

It is only natural to ponder the boundaries of power electronics. In other words at what power levels does it commence and perhaps end? A partial answer resides in the allusion to *heat removal*. In a general way we may assume that we are dealing with power electronics when heat removal merits consideration. In practice it is found that this situation prevails whenever a solid-state device must dissipate 1 watt or more. At 10 watts there is little question of such need. Somewhat arbitrarily we may say that the beginnings of power electronics are in clear evidence at the 3-watt power dissipation level. With regard to an upper limit, this is best left undefined because history suggests the hedged statement that the sky is the limit.

USE OF SOLID-STATE POWER ELECTRONICS

It sometimes happens that a meaningful answer to a question is provided by *another* question. It is appropriate to the subject of solid-state power electronics that we might ask where it is *not* applicable. Certainly such a query would invoke but a small fraction of the applications listed in Table 1-1. The implication is that solid-state power electronics has an almost universal involvement in modern technology. Milestones on the path of electronics progress were marked by such advents as microwaves, television, semiconductors, and computers. And it appears abundantly clear that the present era largely involves the domain of solid-state power electronics.

UNIQUENESS OF SOLID-STATE POWER ELECTRONICS

Much of the answer to such a query is surely well known to the reader. All of the desirable characteristics of semiconductor devices apply. That is, the small size, long life, low cost, easy application, low maintenance, and physical ruggedness of semiconductor devices are as attractive for power electronics as for other applications. But the truly compelling feature of solid-state power devices is the comparative ease with which heat removal can be implemented. This is due to several factors. First, solid-state devices tend to have higher electrical efficiency than other electronic control devices so that less heat is generated within the device itself. (Tubes, in contrast, have high plate-cathode voltage drops. Their "heaters" are well named, for much more energy is converted to heat

TABLE 1-1.   Some Application Areas of Solid-State Power Electronics

ADVERTISING	MAGNETIC RECORDING
AIR CONDITIONING	MOTOR CONTROL
AIRCRAFT	MOVIE PROJECTORS
ALARMS	MOVING LIGHTS
AMPLIFIERS	NEON SIGNS
APPLIANCES	NIXIE DRIVERS
AUTOMOBILES	OVEN CONTROLS
BATTERY CHARGERS	PROJECTORS
BLENDERS	PHONOGRAPHS
BLOWERS	PHOTOCOPIERS
BOILERS	PHOTOGRAPHY
BUOY LIGHTS	POWER LINE CONDITIONERS
BURGLAR ALARMS	POWER SUPPLIES
CHEMICAL PROCESSING	RADAR
CLOTHES DRYERS	RADIO TRANSMITTERS
COMMUNICATIONS	RANGES
CONVERTERS	REFRIGERATORS
CRANES	REGULATORS
DIMMERS	RELAY REPLACEMENT
DISPLAYS	SANDERS
DOOR OPENERS	SEWING MACHINES
ELECTRIC BLANKETS	SERVO SYSTEMS
ELECTRIC VEHICLES	SOLAR POWER
ELEVATORS	SOLENOID DRIVERS
ELECTROLUMINESCENT PANELS	SONAR
ELECTROMAGNETS	SPACE VEHICLES
ELECTRONIC IGNITION	SWEEP CIRCUITS
ELECTROSTATIC PRECIPITATORS	SWITCHING
FANS	TELESCOPE CONTROL
FLASHERS	TELEVISION CIRCUITS
FLOOR POLISHERS	TEMPERATURE CONTROL
FOOD MIXERS	THYRATRON REPLACEMENT
FREQUENCY CONVERTERS	TIMERS
GAMES	TOOLS
GARAGE DOOR OPENERS	TOYS
GAS APPLIANCES	TRAFFIC LIGHT CONTROL
GRINDERS	TRAINS
HAND TOOLS	ULTRASONIC PRODUCTS
HEAT CONTROL	UNINTERRUPTIBLE POWER SUPPLIES
IGNITION	UTILITY SYSTEMS
INDUCTION HEATING	VACUUM CLEANERS
INDUSTRIAL PROCESS CONTROL	VEHICLES
INVERTERS	VENDING MACHINES
LASERS	VENTILATING EQUIPMENT
LATCHING RELAYS	WASHING MACHINES
LATHES	WELDERS
LIGHT CONTROL	WIND POWER
MACHINE TOOLS	XEROGRAPHY
MAGNETS	ZERO-VOLTAGE SWITCHING

than goes into thermionic emission of electrons.) Second, it is generally more practical to devise effective thermal hardware for a compact device fabricated of solid materials in intimate contact than it is for a large and perhaps irregularly shaped device in which the heat-generating elements are not thermally accessible. Third, the operating temperature of silicon devices can be quite high—the early notion of a semiconductor device as being vulnerable to a few tens of degrees Celsius above room temperature is no longer valid.

In addition to the thermal and other features of semiconductor devices, we find that the performance of solid-state power devices is admirable in other respects. Thus, one is not hard pressed to discover such devices with excellent linearity or with speedy response. And although the law of trade-offs cannot be altogether evaded, its assertion is often quite benign for many practical implementations. For example, a transistor that can switch 500 volts at a 20-kHz rate and provide several amperes from a regulated power supply with a switching efficiency of over 90 percent is a mundane rather than an exotic item.

It would be presumptuous to imply that all older control devices (Fig. 1-1) have been rendered obsolete. It is true, however, that solid-state manipulation of power exhibits superiority in such a wide spectrum of applications that it merits study as a unique control technique.

BECOMING PROFICIENT WITH
SOLID-STATE POWER ELECTRONICS

To be well-versed in the general concepts and practices underlying "ordinary" electronics is an excellent foundation for gaining proficiency with solid-state power electronics. However, power electronics differs from the more ordinary phase of the electronics art in that it requires stricter compliance with the basic postulates of electrical engineering. Such a statement may appear trite in the sense that *all* electronic phenomena are governed by the fundamental laws laid down by the physical sciences. In the world of practical experience, however, one often gets by quite well in dealing with ordinary electronic circuitry by paying scant heed to many things that assume importance when power levels are high. In power electronics one is much less likely to enjoy the luxury of another "cut and try" remedy for unsatisfactory results. Whereas the penalty for violation of some basic electrical law may have previously been a glowing plate in a vacuum tube, semiconductor devices are rel-

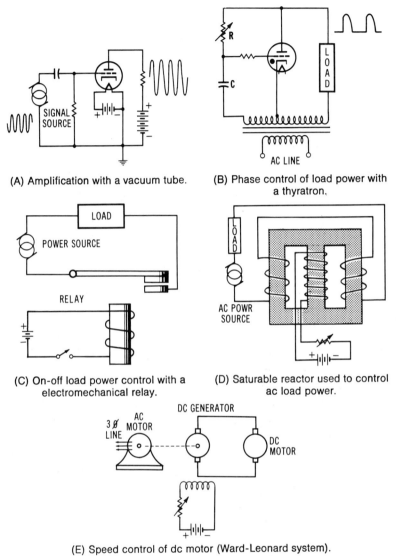

(A) Amplification with a vacuum tube.

(B) Phase control of load power with a thyratron.

(C) On-off load power control with a electromechanical relay.

(D) Saturable reactor used to control ac load power.

(E) Speed control of dc motor (Ward-Leonard system).

Fig. 1-1. Some nonsolid-state control methods.

atively unforgiving. Not only are they vulnerable to destruction when their ratings are exceeded, but they often become involved in a chain reaction in which a number of other devices and components also suffer destruction.

Some of the matters that assume importance in power electronics are power factor, transients, the measurement and consequences of various waveforms, polyphase circuits, duty cycle, energy storage, switching phenomena, and the behavior of low-Q resonance. Although these topics also pertain to ordinary electronics, they merit a more focused emphasis when working in the domain of power electronics. They will be discussed accordingly with this relevancy in mind.

Interpretation of Measurement and Specifications

Many misunderstandings and malperforming circuits occur in power electronics because of the misinterpretation of measurement data. For example, a large switching-type power supply will be ordered for a digital system. It is found that although the supply provides its rated voltage and current and provides excellent regulation, it causes false triggering of sensitive logic circuits. At first, even though the power supply is suspect, the trouble is elusive because ripple and noise from the supply are well within the manufacturer's specifications. Eventually it is discovered that these ratings are specified in terms of their rms value—a perfectly acceptable method per se. However, some of the ripple and noise components are far from sinusoidal waveshape. So, instead of the ratio of peak to rms value being a mere 1.41, as with sine waves, this ratio may be 5 to 1, 10 to 1, or even greater, as tends to be the case with switching spikes. Unfortunately logic circuits tend to be trigger happy—they respond to *peak,* rather than rms, levels. Inasmuch as such transients are not always easily eliminated with filtering and bypassing techniques, it is always desirable to anticipate such possibilities.

Another consequence of such short-duration transients is the actual destruction of semiconductor devices. One might contend that the energy content of very narrow spikes is negligibly low (as proven by the rms value). However, they may easily prove to be the "straw that breaks the camel's back." This does not imply that the destroyed devices were necessarily operating without adequate safety margin. Rather it is a manifestation of the fact that these devices tend to be voltage sensitive in the sense that a momentary penetration of their SOA (safe operating area) boundaries suffices to trigger catastrophic destruction. And such destruction is often possible even after the designer has taken suitable precautions to protect against the "worst possible" overvoltage that could occur under ordinary operating conditions. The proper protection against destructive transients is, of course, an awareness of the various measurement parameters of voltage and current levels, particularly as they apply to nonsinusoidal waves.

In the example just cited, the ideal situation would prevail if the vendor would provide specifications in *both* rms and peak values. Additionally the user should be informed of the *type* of instrumentation employed to provide the measurements. Among other things, instruments, such as oscilloscopes, with inadequate bandwidth would not yield correct indications of narrow spikes. Therefore, it behooves the practitioners to remain alert to technical facts of life and to make their own interpretations.

The Dangers of Accepting Measurements at Face Value

Many of the voltage measurements made in ordinary electronics pertain to direct current (dc) levels and to sine waves. It is only too easy to fall into the habit of quickly grabbing any measuring instrument available in order to make these measurements. Fortunately, dire results from such spontaneous action seem to be relatively infrequent. In solid-state power electronics, however, it is prudent to exercise more caution in arriving at conclusions from measurements. Otherwise, catastrophic destruction may result. At best, undesirable performance and unrewarding analysis are all too likely.

Although it is always necessary to select meters in terms of their sensitivity, impedance, and maximum range, it is often forgotten that the meter must also be related to the *waveform* being evaluated. In this respect meters have three important features, one being the waveform to which they inherently respond and another being the kind of waveform values that their *scales* are calibrated to indicate. For example, a meter may actually detect the peak-to-peak excursions of a waveform, but the scale may be calibrated to provide readout of rms values. But *even this information* may not suffice to obtain accurate readings, for, third, we must also know for what type of waveform the scale calibration was made. If, as is often the case, the rms scale calibration was made for sine waves, the meter may yield erroneous readings when attempts are made to obtain the rms value of a low-duty-cycle rectangular waveform.

Considering that such measurement values as peak-to-peak, rms, and average are geometrically related parameters, one cannot go wrong by using the *oscilloscope* as a voltmeter and making the necessary computations to convert the displayed peak or peak-to-peak values to rms or average values. However, a meter is often more convenient to use and is capable of high accuracy if we know the nature of the waveform and the behavior of the meter. Table 1-2 depicts the situations prevailing for various meters and for commonly encountered waveforms in solid-state power electronics. Wherever we see the factor of unity (1.000), the meter readout is *true*. Other factors are multipliers needed to correct the observed reading to true value. Interestingly some combinations of meter types and waveforms produce *no* true readings for any of the four geometrical parameters.

Thyristor Waveforms

The waveforms shown in Fig. 1-2 are those associated with a simple triac phase-control circuit and offer additional opportunities for erroneous interpretation of measurements. If one studies the rms value of the currents, it is obvious that the meter used for making the measurements must be responsive to the rms value of very nonsinusoidal waves. Thermocouple meters, specially designed electronic meters, and some digital types can satisfy this requirement most readily. The iron-vane meter and the dynamometer type may also qualify, but both of these meters tend to depart from accuracy with certain nonsinusoidal inputs. It is always best to learn about the response of a meter rather than assume that it is immune to the effects of waveform distortion. (For example, the fixed and movable coils of the dynamometer type sometimes provide sufficiently different reactance to the harmonics of a nonsinusoidal input to result in an inaccurate indication.)

Further investigation of the data presented in Fig. 1-2 reveals features contrary to common sense. For example, note that the power factor of the load is unity in all instances as one would expect from a purely resistive load. It is natural to suppose that power factors less than unity are the result of phase displacement between voltage and current *only* because of inductance or capacitance. Yet we see that the *line* power factor rapidly departs from unity as the conduction waveform is reduced from 180 degrees. Inasmuch as we are dealing with a simple series circuit, it may appear strange that the power factor is *different* on the two sides of the triac. There is a difference between these two circuit sections, however. Specifically, on the line side the distorted current wave is

TABLE 1-2. Waveform Conversion Factors for Various Types of Voltmeters

WAVEFORM		VOLTMETER TYPES					
		Peak-to-Peak Responding OSCILLOSCOPE ELECTRONIC-METER Peak-to-Peak Scale	Peak Responding ELECTRONIC-METER Peak Scale	Peak Responding ELECTRONIC-METER RMS Scale (Calibrated for sine waves) Also some DIGITAL METERS (not common)	RMS Responding IRON VANE THERMOCOUPLE DYNAMOMETER RMS ELECTRONIC-METER RMS Scales (Calibrated for sine waves) Also some DIGITAL METERS (newer types)	Average Responding RECTIFIER-METER RMS Scale (Calibrated for sine waves) Also most older DIGITAL METERS	Average Responding RECTIFIER-METER Average Scale (Calibrated for sine waves)
SINE:	pk-pk	1.000	2.000	2.828	2.828	2.828	3.140
	0 pk	0.500	1.000	1.414	1.414	1.414	1.570
	rms	0.353	0.707	1.000	1.000	1.000	1.111
	avg	0.318	0.637	0.900	0.900	0.900	1.000
RECTIFIED SINE: (FULL WAVE)	pk-pk	1.000	1.000	1.414	1.414	1.414	1.570
	0 pk	1.000	1.000	1.414	1.414	1.414	1.570
	rms	0.707	0.707	1.000	1.000	1.000	1.111
	avg	0.637	0.637	0.900	0.900	0.900	1.000
RECTIFIED SINE: (HALF WAVE)	pk-pk	1.000	1.000	1.414	2.000	2.828	3.140
	0 pk	1.000	1.000	1.414	2.000	2.828	3.140
	rms	0.500	0.500	0.707	1.000	1.414	1.570
	avg	0.318	0.318	0.450	0.637	0.900	1.000

continued

TABLE 1-2. *continued*

Waveform		Col 1	Col 2	Col 3	Col 4	Col 5	Col 6
SQUARE 	pk-pk	1.000	2.000	2.828	2.000	1.800	2.000
	0 pk	0.500	1.000	1.414	1.000	0.900	1.000
	rms	0.500	1.000	1.414	1.000	0.900	1.000
	avg	0.500	1.000	1.414	1.000	0.900	1.000
RECTIFIED SQUARE 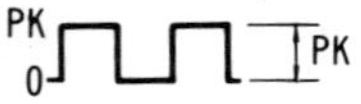	pk-pk	1.000	1.000	1.414	1.414	1.800	2.000
	0 pk	1.000	1.000	1.414	1.414	1.800	2.000
	rms	0.707	0.707	1.000	1.000	1.272	1.414
	avg	0.500	0.500	0.707	0.707	0.900	1.000
RECTANGULAR PULSE 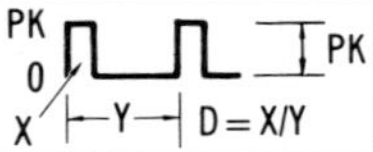	pk-pk	1.000	1.000	1.414	$1/D^{1/2}$	$0.9/D$	$1/D$
	0 pk	1.000	1.000	1.414	$1/D^{1/2}$	$0.9/D$	$1/D$
	rms	$D^{1/2}$	$D^{1/2}$	$1.414\ D^{1/2}$	1.000	$0.9/D^{1/2}$	$1/D^{1/2}$
	avg	D	D	$1.414\ D$	$D^{1/2}$	$0.9\ D$	1.000
TRIANGLE 	pk-pk	1.000	2.000	2.828	3.464	3.600	4.000
	0 pk	0.500	1.000	1.414	1.732	1.800	2.000
	rms	0.289	0.577	0.816	1.000	1.038	1.153
	avg	0.250	0.500	0.707	0.867	0.900	1.000

SQUARE: PK, PK-PK

RECTIFIED SQUARE: PK

RECTANGULAR PULSE: PK, $X \vdash Y \dashv D = X/Y$

TRIANGLE: PK, PK-PK

compared to the *sine* wave of voltage from the line; on the load side the voltage and current waves are always *identical*. On the line side, only the *fundamental* of the current wave can be said to be in phase with the line voltage. Many harmonics of the current wave are not in phase with the line voltage—it is as if these phase differences were caused by a physical reactance.

The practical consequence of such low-power-factor operation is that the circuit is much less efficient at low conduction angles than one would infer from the near-ideal switching behavior of the triac. For a given amount of real power delivered to the load, the electric utility must supply more current than would be necessary at unity power factor on the line.

Although meters tend to be more convenient than oscilloscopes, it is obvious that the oscilloscope is indispensable as a means of qualitatively ascertaining the nature of the waveform being measured by a meter. With regard to digital meters, many types have been marketed. There appears to be a trend toward rms-responding types, and some of these have the desirable feature that the response is accurate for a wide variety of waveforms. One thing to keep in mind when using these instruments is their input impedance, which in some instances is low enough to upset certain circuits. Whether digital or analog, a meter that is capable of providing an accurate readout *regardless* of input waveform is exceedingly useful for the measurements encountered in solid-state power electronics.

Also appearing on the market are various "smart" oscilloscopes. These instruments are associated with calculator ICs or microprocessors in such a way that a digital readout of the various wave values is quickly and conveniently presented in addition to the conventional waveform

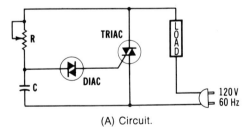

(A) Circuit.

Fig. 1-2. Simple triac phase-control circuit and associated waveforms.

LINE VOLTAGE WAVEFORM (SINE WAVES)	CURRENT CONDUCTION ANGLE IN DEGREES	LINE AND LOAD CURRENT WAVEFORM. (LOAD VOLTAGE WAVEFORM.)	EFFECTIVE CURRENT (RMS VALUE)	EFFECTIVE VOLT-AMPERES	ACTUAL POWER IN LOAD RESISTANCE	POWER FACTOR AT AC LINE.	POWER FACTOR AT LOAD
1	2	3	4	5	6	7	8
 Em	180	 Im	0.707 Im	0.500 EmIm	0.500 EmIm	1.000	1.000
	150		0.697 Im	0.493 EmIm	0.486 EmIm	0.985	1.000
	120		0.634 Im	0.448 EmIm	0.402 EmIm	0.897	1.000
	90	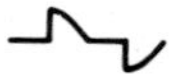	0.500 Im	0.354 EmIm	0.250 EmIm	0.707	1.000
	60		0.313 Im	0.221 EmIm	0.098 EmIm	0.443	1.000
	30		0.120 Im	0.085 EmIm	0.014 EmIm	0.170	1.000

(B) Waveforms.

Fig. 1-2. *continued*

patterns. Also, rates of rise and decay are speedily and accurately indicated. It is anticipated that these instruments will prove very useful
in solid-state power electronics.

In electrical power work it is customary to use wattmeters and
power-factor meters to evaluate the performance of motors, heaters,
and other loads. However, when the rms current to a load is controlled
by a thyristor, the use of these instruments becomes suspect. Again, it
is wise to establish the response characteristics of such meters before
making measurements in silicon controlled rectifier (SCR) or triac circuits. Often, corrective curves can be plotted so that departures from
accuracy can be remedied as readings are recorded or afterward.

The examples given for triac operation also apply to back-to-back
operation of SCRs. At the present, SCRs have greater power-handling
capability than do triacs. Whereas triacs may be found in light dimmers
and in control circuits for hand tools, SCRs will be encountered in
applications involving large industrial motors or in traction vehicles.
Single SCR circuits behave similarly, except that only half-wave power
is delivered to the load. Here the chances of grossly erroneous meter
readings are even greater, for the current wave is mathematically nonsinusoidal even at current conduction angles of 180 degrees.

THE BASIC TECHNIQUES FOR
MANIPULATING ELECTRICAL POWER

Our investigation of solid-state power electronics will be considerably
aided if we first consider the ways in which solid-state devices may be
used to process or control electrical power. The number of different
operating modes that have come into use will likely prove surprising. It
is significant to note that, although some of these techniques represent
long-known concepts, they have only recently been implemented in
practice.

On-Off Switching

Nothing is implied here beyond merely turning on and off the power
applied to a load. Either dc or alternating current (ac) may be applied
and interrupted. A switch in the ac power line may be used, the solid-
state power device may be connected and disconnected from the power

source, or the trigger or control electrode of the power device may be actuated to render the device conductive or nonconductive. This appears simple enough from one's experience with low-power equipment. But in every circuit there is at least *stray* inductance and capacitance. When current is made or interrupted, a "stray" emf will be generated according to the relationship, $e = L \, di/dt$. Also, an "unintended" current, i, will flow according to the relationship, $i = C \, dv/dt$. These switch-produced voltages and currents are known as *transients* or *surges*. When turning on and off large blocks of power, it is usually *not* sufficient merely to arrange for making or breaking contacts or for abruptly making a solid-state device conductive or nonconductive. Even a one-time transient can be in the nature of a destructive arc or a voltage spike greatly exceeding the "dielectric" strength of a control device or a current surge exceeding the maximum current rating of a device.

Somewhat more insight is provided by contemplation of the energy equations that are operative at all times in electrical circuits. They are

$$W = \frac{I^2 L}{2}$$

or

$$W = \frac{E^2}{2C}$$

where

W is in watt-seconds or joules
I is in amperes
E is in volts
L is the inductance in henrys
C is the capacitance in farads

These equations tell us that even though L and C may exist primarily due to wiring and the proximity of circuit board components, the stored energy may be appreciable if E or I is high. For example, the energy stored in the magnetic field of a given conductor is one million times greater if the steady-state current is 10 amperes, rather than 10 milliamperes.

It happens that there is not much we can do about the stray energy stored in the magnetic and electric fields of a power circuit, once we have arranged for a reasonable wiring layout. And to make matters worse, energy storage will often be much higher in the *physical* capacitors and inductors in most circuits. For example, if a simple rectifier-filter system is turned on, the filter capacitor initially appears as a short circuit to the applied dc. As the capacitor charges, this condition is alleviated. However, at the first instant the rectifier elements are subjected to a tremendous current surge. In power equipment such a surge can easily destroy these rectifiers. At best the designer may be forced to provide expensive rectifiers that have continuous operation ratings greatly in excess to that needed.

Although we have been discussing the simple *one-time* turn on and turn off of power equipment, this discussion also applies to numerous techniques wherein power is switched periodically, such as inverters, switching-type regulators, and phase-controlled thyristor circuits. And because of energy storage and switching transients, almost all solid-state power circuits make use of *suppression* techniques. That is, the energy content of the transients is reduced or their energy is largely absorbed in some device designed for this purpose. One way to reduce the energy content of switching transients is to slow the rise and fall times of the switching process. Unfortunately, although this is readily accomplished with the one-time turn on or turn off of equipment, it can be applied only in a limited way to any circuit involving repetitive switching. In such circuits the efficiency of the switch suffers if too much time is consumed in turn on and turn off. (Any finite time spent by a switch in getting from one state to another is a time of power dissipation in the switch itself.) Any practical design must balance these conflicting needs for rapid and limited-rate transition times in a switching device.

Another reason for suppressing switching transients is that they are a source of *interference* to other circuits and systems. Much of the energy present in sharp switching transients radiates and is particularly difficult to combat once it finds its way into sensitive circuitry, such as logic systems. Do not be surprised to find common transient-suppression techniques used in otherwise diverse solid-state power circuits.

Power Control via Switching Techniques

It is surprising how many different ways there are to control the amount of electrical power consumed by a load. If the power source is dc, the most obvious control element is some form of rheostat or potentiometer. This is entirely acceptable and has much to commend it. For example, such control is easily made, smooth, continuous, wide range, and may readily be made proportional to a quantity such as shaft rotation or programming voltage. However, one must be reconciled to live with the "throwaway" power that must be dissipated in the control element. Unfortunately in power electronics it often proves neither economic nor otherwise permissible to operate at the inherently low efficiencies provided by such resistive control. Appropriately enough, such a control technique, whether implemented by a physical rheostat or the latest monolithic Darlington transistor, is known as *dissipative* control. Because of the problems of cost and heat removal associated with dissipative control, more efficient power-control methods are used in power electronics.

With regard to the control of dc power, it often proves profitable to convert it to ac, which is subject to more efficient control techniques, then convert back to dc via rectification and filtering. Another much-used technique is to "chop" the dc and pass the rough waveform through a low-pass filter before presentation to the load. The power delivered to the load can be controlled by means of either of two chopping or switching modes. The pulsed dc can be duty-cycle modulated with the chopping frequency remaining constant. This is generally referred to as *pulse-width modulation.* Or the pulse width can be maintained constant while the frequency or repetition rate is varied. In both cases the average voltage, *Vo,* presented to the load is given by the equation:

$$V_o = \left(\frac{T_{on} - T_{off}}{T_{on}} \right) V_{in}$$

where

V_o is the average voltage presented to the load
T_{on} is the on time
T_{off} is the off time
V_{in} is the source voltage

The waveforms pertaining to these control methods are illustrated in Fig. 1-3. The low-pass filter can be thought of as an averaging or integrating circuit in addition to an attenuator of the switching-frequency pulses

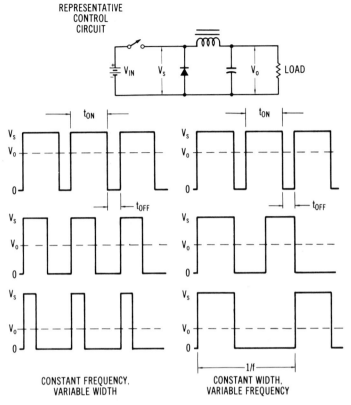

Fig. 1-3. Switching waveforms in dc power-control systems.

and its harmonics. The freewheeling diode enables load current to be supplied during T_{off}. The source of this current is the energy stored in the magnetic field of the inductor.

The beauty of the *ideal* switching device is that it dissipates *no* power when in the on state because it displays zero resistance; likewise, it dissipates *no* power in the off state because its resistance is infinite. Actual solid-state switching devices do not yield such 100 percent control efficiency. They exhibit resistance when on, and a small, but significant, leakage current may flow when they are turned off. Moreover, these devices also have dissipative losses because they can turn neither on nor off in zero time. Although their rise and fall times may be small, the losses can be appreciable if these transition times are comparable to the on time. Nonetheless power control by these switching techniques is capable of much greater efficiency than can be attained with dissipative control.

A more detailed view of the switching waveforms is illustrated in Fig. 1-4. Here, we see that the ideal "pure" dc is not achieved for V_o. Rather there is a superimposed ripple component. The operation of the inductor is also shown in Fig. 1-4. Note that the inductor voltage makes positive and negative excursions around the average value, V_L. The average dc value of the inductor voltage is zero. The inductor current is essentially the load current with a superimposed sawtooth. Because of the freewheeling diode, the inductor delivers load current not only when the switch is closed but also when it is open. These switching waveforms pertain to both the variable width and the variable frequency modes of operation. The output ripple voltage, although a smooth wave,

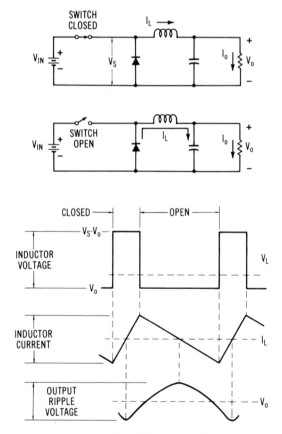

Fig. 1-4.   A more detailed view of switching waveforms.

is not sinusoidal. In practical dc switching supplies, it is not easy to reduce the output ripple to an extremely small value because the electrolytic filter capacitors have appreciable equivalent series resistance. And increasing the size of the inductor adversely affects speed of response in a regulated supply, to say nothing of cost, weight, and packaging problems.

A particularly advantageous scheme of dc power control is to rectify and filter the incoming ac from the power line, control the resulting dc by means of a switching device, and then pass the chopped waveform through a low-pass filter to the load. It often happens that a 60-hertz power transformer is not needed because this type of switching control can operate efficiently at a low duty cycle in order to deliver a dc load voltage considerably below line voltage. (If this is attempted with a dissipative control, such as the common series-pass transistor in regulated power supplies, the power lost in the rheostatlike control device may easily exceed the power delivered to the load.)

A somewhat more sophisticated version of the preceding system uses the rectified and filtered ac line power to operate a dc to ac inverter, which has its ac duty-cycle modulated or programmed. Of course, the ac from the inverter must then be rectified and filtered again so that the load may have controlled dc power applied to it. At first it might appear that this is a roundabout approach that unnecessarily requires extra components. Actually the savings in size, weight, and cost that can be effected by this dc power control method are considerable. This comes about by virtue of the *high* operating frequency of the inverter. This operating frequency may be 1 to 50 kilohertz, or even higher. A good balance among many contradictory factors involving various frequency-dependent losses, together with the costs of components, dictates a switching rate in the vicinity of 20 to 25 kilohertz. This is above the general audible threshold, sometimes an important consideration in power-control systems because transformers and chokes can radiate annoying levels of accoustic energy. These relatively high frequencies require only minuscule core components. The elimination of 60-hertz magnetics is then the compelling feature of such a system—the few extra parts are trivial contrasted to the gains achieved by dispensing with the *bulky* and expensive 60-hertz transformer.

Most dc power-control systems, whether they utilize dissipative or switching techniques, are generally closed-loop arrangements that provide either constant voltage or constant current to the load regardless of wide variations in ac line voltage, or in the load itself. Such automatic stabilizing systems are collectively known as *regulated power supplies*. They constitute one of the most important types of circuits encountered in solid-state power electronics.

AC Power Control—The Phase-Control Method

It is obvious that ac power can be controlled by rheostatlike devices in similar fashion to the dissipative control of dc power. And likewise one must then be prepared to cope with low efficiency, high cost, and awkward thermal problems. In principle, ac power may be controlled with high efficiency and low dissipation by means of variable reactance. Practical implementations have employed such devices as magnetic amplifiers and saturable chokes. Thus far, no practical utilization of variable capacitance has emerged because the physical size of such a device at power-line to supersonic frequencies would rule out its consideration.

One of the most popular power-control methods is the phase control of the ac sine wave by means of thyristors. Fig. 1-5 illustrates both half-wave and full-wave voltage waveforms in which the rms power delivered to the load is governed by the timing of the gate turn-on trigger applied to the thyristor. The reason underlying the high efficiency of this type of control is that its operation is essentially that of a *switch*— it is either fully on or completely off (ideally). The half-wave voltage

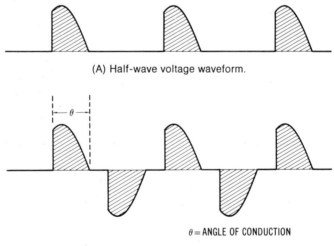

(A) Half-wave voltage waveform.

θ = ANGLE OF CONDUCTION

(B) Full-wave voltage waveform.

Fig. 1-5. Load-voltage waveforms in phase-controlled power systems.

shown in Fig. 1-5A is typical of single SCR control of the load. The full-wave scheme (Fig. 1-5B) can be implemented either by a single triac or by back-to-back SCRs. SCRs with power-handling capabilities greatly exceeding other solid-state devices have become available at nominal cost. The voltage and current capabilities of triacs are constantly being improved. These considerations tend to override the disadvantages inherent in the very nonsinusoidal waveforms produced. These waveforms produce electrical noise, are difficult to measure with commonly available instrumentation, and can result in operation at low power factor.

Pulse-Modulation Method of AC Power Control

The load voltage waveforms associated with two pulse-modulation techniques are shown in Fig. 1-6. These waveforms are usually produced by inverters. In Fig. 1-6A equal-duration pulses of varying amplitudes are appropriately timed to approximate the general contour of a sine wave. If each of the wave building-blocks is made proportionately greater or smaller in amplitude, the rms value of the composite wave is varied, but its essential shape remains unaffected. If such a waveform is "strained" through a low-pass filter, a reasonably good sine wave may result. In practice the windings of a motor may suffice for this purpose.

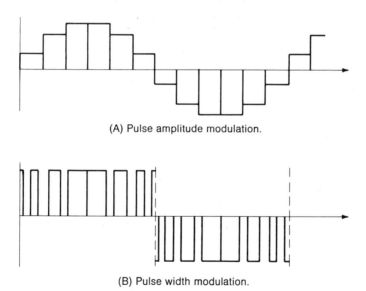

(A) Pulse amplitude modulation.

(B) Pulse width modulation.

Fig. 1-6.   Load-voltage waveforms in pulse-controlled power systems.

The simulated sine wave depicted in Fig. 1-6B is pulse-width modulated—the amplitude remains constant, but the width of the pulse is apportioned so that the distribution of power within the composite wave is approximately sinusoidal. Power control may be affected by varying the widths of the pulses but at the same time maintaining their *relative* durations. As with the waveform in Fig. 1-6A, a low-pass filter may be used to produce an acceptable sine wave.

These "fabricated" ac sine waves are often found in power-control systems in which the control parameter is frequency rather than rms value. Special considerations then apply to the widths of the pulses in both waveforms. However, this is easily taken care of in low-power logic circuitry. These wave-construction techniques also are useful in designing three-phase power equipment. Again, the phase displacements between the waveforms are accomplished in low-level logic circuits. Required flexibilities in control functions are more readily attained with these fabricated waveforms than by resorting to "brute-force" analog methods. For example, a speed-control system for three-phase induction motors requires that the applied motor voltages go up and down with the impressed frequency. In addition there are power-output circuits configured about both transistors and thyristors. (Although their power-handling abilities are relatively limited, both *gate turn-off SCRs* and *power MOSFETs* merit consideration for such service.)

Burst-Modulation Method of AC Power Control

It had long been apparent that for certain types of loads, such as heating elements, it was possible to control average ac power in yet another way. If varying lengths of ac wavetrains are metered out to a load, an *average* power level dissipated in the load will be less than would be experienced if the ac cycles were continuously applied. And the greater the length of time between successive wavetrains, the less will be the average load power. However, for this basic technique to achieve maximal usefulness, a certain refinement is desirable. Specifically, only complete sine waves should be metered out to the load. Otherwise one can anticipate considerable rfi and emi. Any time a sine wave is chopped so that there are fractional cycles between voltage zero crossings, the wavetrains are nonsinusoidal from a mathematical viewpoint. This alone can result in the production of many higher harmonics capable of creating interference. Additionally the application of voltage at nonzero levels is tantamount to high *dv/dt* rates, which further enriches that portion of the electromagnetic spectrum contributing to electrical noise.

If on the other hand this control technique is implemented under the condition that only wavetrains comprising *complete cycles* are metered out to the load, there will be no rfi, emi, or harmonics produced. Fig. 1-7 illustrates such power control. Note that these load voltage waveforms commence and cease only at *zero crossings*. Of course this control scheme is best adapted to "sluggish" loads—those that would not be likely to respond with intermittent operation. This restricts its use mainly to heaters, but under some circumstances motor speed control is also feasible.

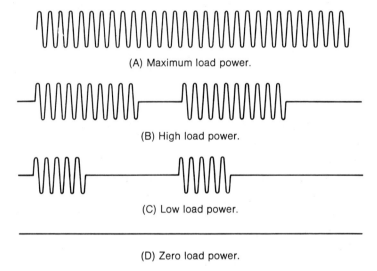

(A) Maximum load power.

(B) High load power.

(C) Low load power.

(D) Zero load power.

Fig. 1-7. Load-voltage waveforms in burst-modulation-controlled power systems.

EVALUATION OF POWER AMPLIFIERS

Many solid-state devices are used as power amplifiers, and quite a few circuit techniques are available for accomplishing power amplification. Accordingly it would be useful if the basic performance of power amplifiers could be described by a standardized figure of merit. That is, we should be able to visualize any device or any circuit technique capable of developing amplification at high power levels as a "black box" possessing a meaningful quality as a power-level booster. In the technical literature we find that transistors are commonly specified in terms of their current gain at ordinary frequencies and by their power gain in the uhf and microwave portion of the spectrum. However, the amplifying ability of power FETs is mainly specified by their transconductance. When either device is used in a pulse-width-modulated, or Class D, amplifier, there seems to be no set rule for relating the output to the input.

Actually all amplifiers are *power* amplifiers even though we often term them otherwise, such as *voltage* or *current amplifiers*. (If we are truly interested in voltage or current gain *only*, the ordinary *transformer* can provide such "amplification.") An appropriate figure of merit must involve *both* voltage and current in such a way that one isn't gained simply at the expense of the other such as in a transformer. The use of the term *power gain* involves practical difficulties for such a device as the power FET, in which input power is negligibly small. Similarly the power FET is not readily described by current gain because of its minute input current. This current would be difficult to measure, and it would not be easy to determine what portion of it was due to leakage effects. Voltage gain of the power FET might lead to some useful results but would fall short of practicability when applied to bipolar transistors. In any event, voltage or current gains must be related to input and output impedances to be truly meaningful for power amplifiers.

It turns out that the concept of *transconductance* in which output current is related to input voltage is applicable to *any* black box serving the function of amplification. See Fig. 1-8. Not only is this figure of merit used with the new power FETs, but manufacturers of bipolar power transistors are providing transconductance data in their specifi-

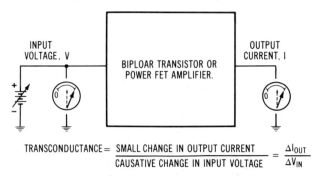

$$\text{TRANSCONDUCTANCE} = \frac{\text{SMALL CHANGE IN OUTPUT CURRENT}}{\text{CAUSATIVE CHANGE IN INPUT VOLTAGE}} = \frac{\Delta I_{OUT}}{\Delta V_{IN}}$$

Fig. 1-8. Transconductance concept suitable for solid-state power amplifiers.

cations to an increasing extent. The transconductance of any device or circuit is given as the ratio of a small change in *output current* to the change in *input voltage* causing the change in output current, while all other operating parameters are maintained constant. Interestingly, whereas anything in excess of 10,000 micromhos was considered a "hot" vacuum tube, power transistors can have transconductance values in the *millions* of micromhos range. From such curves as are shown in Fig. 1-9, the transconductance at an operating point such as P is determined by the equation

$$gm = \frac{\Delta I_{out}}{\Delta V_{in}}$$

where

gm is the transconductance
ΔI_{out} is a small change in output current
ΔV_{in} is a small change in input voltage

Note that the collector voltage must be held constant when applying this concept.*

Actually many designers of transistor circuitry have continued to use transconductance as a heritage from their vacuum-tube days. The transconductance of a transistor can be derived from the relationship, $gm = h_{fe}/h_{ie}$. However, it is more conveniently determined from graphs of collector current versus base-emitter voltage, which power transistor manufacturers are providing with their specification sheets. Although the true mathematical use of transconductance precludes the insertion of a load in the output circuit, many amplifier circuits used in power electronics tend to have about the same transconductance as do the active devices alone. This is because these amplifiers are for the most part not employed as so-called voltage amplifiers, because the loads have relatively low output impedance. Yet another reason why the approximation is usually good is that power transistors and power FETs display "pentode" characteristics. That is, they behave as constant-current generators (output current is about the same with a zero-impedance load or the actual load).

---

*In vacuum tubes, the plate (and screen grid, if present) must be held at a constant voltage. When using FETs, the drain voltage must be held constant.

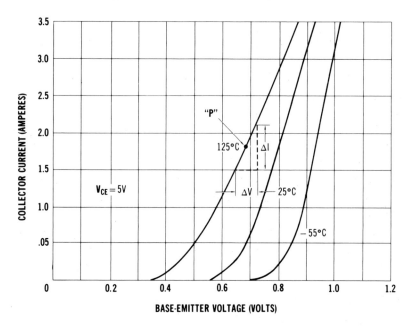

Fig. 1-9. Transconductance characteristics of the Delco DTS-424 power transistor.

Although the ability to visualize any amplifier as a "transconductance stage" provides operational insight, one must pay heed to other factors effecting the practical implementation of a power amplifier. Consider, for example, two Class A power amplifiers displaying the same overall transconductance, but one is designed with bipolar transistors, whereas the other uses FETs. A significant practical difference is that the FET amplifier will consume negligible current from the signal source. Its bipolar transistor counterpart on the other hand must be supplied with substantial signal current. (Although it is easy to say that the FET amplifier has the higher "power gain," this does not argue well for the use of power gain as the standard yardstick for evaluating amplifiers. A little reflection reveals that *any* FET device may be assigned infinite power gain when operating as a Class A amplifier.) Also transconductance often must be used with other relevant parameters such as input capacitance, frequency limits, voltage and current boundaries. But transconductance can be a good place to start in the design or the analysis of a solid-state power amplifier. The transconductance concept can be applied directly to other types of solid-state power amplifiers, such as parametric amplifiers, negative impedance amplifiers, and Class D amplifiers, and to magnetic amplifiers using diodes.

CLASS A AMPLIFIER

The Class A amplifier delivers a proportional output over the entire cycle of input signal. In other words this is a linear amplifier. It is not surprising that the Class A power amplifier was extensively used in audio equipment. It is not popular any longer for higher power audio amplifiers because of its inherently poor efficiency, which approaches 50 percent under ideal conditions. Inasmuch as audio signals have high peak-to-average ratios, the efficiency of the Class A amplifier is quite low in actual sound applications. For a long time this was tolerated because of its excellent linearity. In particular, push-pull Class A power amplifiers tended to cancel even-harmonic distortion. Such audio equipment often produced high-quality reproduction with minimal local or loop feedback. Indeed there remains a number of devout stereo buffs who feel that Class A power amplification is the superior mode for faithful audio reproduction. However, when the consumer demand increased for high power capability, Class A amplification had to give way to more efficient methods. Although many users would have been willing to cope with the higher cost of operation of Class A audio power equipment, the thermal problems proved incompatible with the needs of consumer products. This can be appreciated in light of the fact that a Class A amplifier dissipates half of its available sinusoidal output power while just idling with no input signal.

Class A amplification is still extensively used at lower power levels, such as the driver stages in audio circuitry. Class A amplifiers are often found in industrial applications where efficiency may not be a paramount consideration and where ample space and cooling apparatus can readily be provided. The series-pass device in linear regulators, while not commonly referred to as a Class A amplifier, nonetheless operates in a fashion resembling this amplification mode. As with more conventional Class A power amplifiers, the series-pass device is inefficient because it must dissipate considerable power within itself.

Single-Ended Class A Amplifier

Typical single-ended Class A power amplifiers are shown in Fig. 1-10.

The dynamic operation of these amplifiers is illustrated in Fig. 1-11. Note that the operating point is established to enable optimum utilization of the linear portion of the transfer characteristic curve. The operating point is determined by all three of the biasing resistances, R1, R2, and R3. Unless adequately bypassed over the operating frequency range, emitter resistance R1 also produces current-derived negative feedback. Although such feedback lowers the stage gain, it reduces certain distortion products. Also the input and output impedances of the stage can often be conveniently manipulated by such feedback. This resistance is beneficial in stabilizing the operating point against temperature variations, particularly with germanium transistors, where it can help prevent thermal runaway.

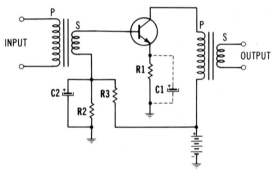

(A) Optimum power gain by use of input impedance-matching transformer.

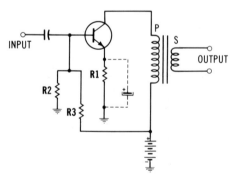

(B) Simpler input coupling arrangement.

Fig. 1-10. Typical Class A power amplifiers.

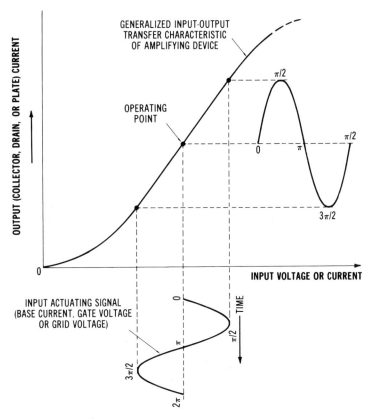

Fig. 1-11.   Representative operating characteristics of all Class A power amplifiers.

The Push-Pull Mode for Class A Operation

A typical push-pull power amplifier circuit suitable for Class A operation is shown in Fig. 1-12. The operating characteristics of such a circuit is illustrated in Fig. 1-13. Note that the operating points of transistors Q1 and Q2 are so situated that the output-current swing involves *more* than just the linear portion of the transfer curves. This depicts one of the compelling features of push-pull operation. Even though each transistor produces considerable second-harmonic distortion, the *composite* output waveform is free of such distortion. This implies that for a given amount of output distortion the push-pull arrangement can yield *more than twice* the output power practically attainable from a single-ended Class A

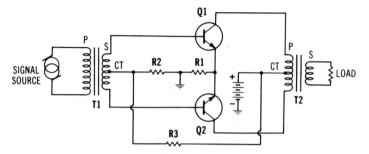

Fig. 1-12. A typical push-pull power amplifier circuit.

amplifier. Inasmuch as low operating efficiency is the principal short-coming of Class A amplification, any expedient that allows greater output to be developed is certainly welcome.

In order to provide clearer insight into the nature of second-harmonic cancellation, the second-harmonic current waveforms in the two halves of the output transformer primary winding are depicted in Fig. 1-13. These second-harmonic components would *not* be observed by probing with an oscilloscope. Only the badly distorted output current waveforms of Q1 and Q2 would be seen on the scope. (However, either a mathematical or a graphical analysis of these current waveforms would yield their constituent harmonic "building blocks." Briefly, the *Fourier theorem* states that any nonsinusoidal wave can be resolved into a number of harmonically related sine waves. The "first harmonic," of the Fourier series of constituent frequencies. Often there is also a zero-frequency or dc component in the series, particularly when *even* harmonics are present. The phase relationships and amplitudes of the various harmonics give rise to any imaginable waveshape.)

In our particular case it can be seen that the second-harmonic waves *cancel* because of their equal amplitude and 180-degree phase relationship. The algebraic addition of the Q1 and Q2 currents take place in the secondary winding of the output transformer. The net result, as seen by the load, is a nice clean sine wave—or at least one devoid of obvious second-harmonic distortion. It should be pointed out that other even-order harmonics are generated by the curvature of the transfer curves of Q1 and Q2. These even-order harmonics, such as the fourth and sixth harmonics, are *also* cancelled in similar fashion to the predominant second harmonic. It should be added that transfer curve non-linearity can also produce odd-order harmonics, the third, fifth, seventh,

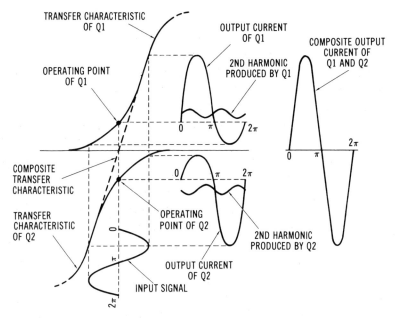

Fig. 1-13.   Second-harmonic cancellation in hard-driven Class A push-pull amplifier.

etc. These odd harmonics distort the smooth sinusoidal shape of the composite output current waveform but tend to produce less troublesome effects than do even harmonics, unless present in unusually high amplitudes.

There are even other advantages of Class A push-pull operation. There is no dc saturation of the output transformer because the center-tap connection causes the direct currents in the two halves of the primary winding to produce opposing fluxes. Not only does this make for a more economical transformer design, but a potential source of distortion products, principally odd harmonics, is eliminated. Bypass capacitors are less critical in the push-pull circuit because signal components need not complete their paths through the dc power supply. And the nature of push-pull operation is such that less imposing demands are made on power-supply filtering. This is because ripple fed to the center-tap of the output transformer tends to cancel. However, there will not be any discrimination against power-supply ripple mixed with, or superimposed on, the input signal.

The advantages of push-pull operation deteriorate rapidly when the two active devices are not identical. It has not always been easy to obtain pairs of power transistors with reasonably matched characteristics. However, if one ascertains that the current gain factors (beta) are matched for corresponding currents and frequencies, good results should be obtained. It is generally not sufficient merely to match current gains at dc or at a low frequency. A wise strategem in power amplifier design is to attend *first* to those matters pertaining to the best linearity and balance that can be had; *then* sufficient negative feedback may be incorporated to further minimize signal distortion. The notion of such feedback as a universal panacea for sloppy amplifier design is a cosmetic approach that never completely disguises the underlying distortion. Many a stereo amplifier fails to convey acoustical integrity to the ear despite decades of flat frequency response, tremendous dynamic power capability, and, of course, multiple decibels of feedback.

CLASS B AMPLIFIER

The Class B amplifier delivers output current and power only over one-half cycle (180 degrees) of an applied sine-wave input signal. Inasmuch as such operation is suggestive of rectification, it is obvious that a single-ended Class B device would produce extremely high distortion in audio applications. However, the push-pull configuration (Fig. 1-14) solves this problem. This is because only one device is conductive at any given time—the other device is cut off until polarity reversal of the input signal alternates the conductive state of the two devices. By the same token, when no input signal is applied, the idling current of the push-pull Class B amplifier is zero. As a result, one would anticipate higher operating efficiency than is available from Class A amplifiers. It turns out that the Class B amplifier has a theoretical maximum efficiency of 78 percent. Such efficiency prevails only when a Class B output stage is driven to its full rated output. Therefore, the operating efficiency with audio signals is generally much lower. Nonetheless, the average efficiency tends to be appreciably greater than is forthcoming from the Class A amplifier operating under similar signal and output conditions. And this is despite the fact that the input-power requirements of most devices operated in Class B is greater than for Class A operation.

Class B power amplification is commonly used in servo amplifiers and in other "brute-force" tasks devoted to power gain. Paradoxically

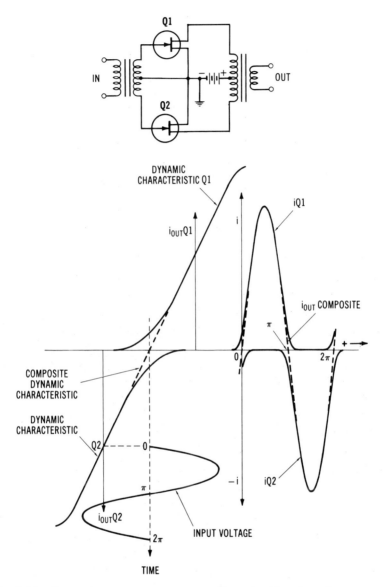

Fig. 1-14. Power devices operating essentially in the Class B mode.

Class B is not the predominant mode encountered in audio work—at least, not in its "pure" form. This is because most devices are not linear at low-power operation levels, which gives rise to a switching transition known as *crossover distortion*. The best way to avoid the most detrimental effects of crossover distortion is to blend in a small amount of Class A operation. That is, although the power amplifier is essentially Class B in the sense that its average dc current swings with the signal, its quiescent output current is relatively low so that the high average dissipation that characterizes Class A operation does not exist. At the same time the switching transition from one of the push-pull devices to the other is smoothed out and crossover distortion is greatly reduced. The significant feature of the Class B push-pull amplifier is that the full waveform is reproduced in the output circuit even though each device is active for only one-half of the input wave.

COMPLEMENTARY-SYMMETRY AMPLIFIERS

The concept of complementary symmetry has been applied ever since both npn and pnp transistors became available. Complementary symmetry allows simple push-pull amplifier stages to be designed. Inasmuch as there is no vacuum-tube version of the pnp transistor available, this interesting circuit certainly could not have been implemented before semiconductor technology. As a matter of fact, until recently the serious deployment of the technique was hampered by the lack of suitable pnp power transistors to form matched pairs with high-quality silicon npn types. Not only has this situation been remedied, but complementary pairs of power FET devices are also being introduced. The technique of complementary symmetry is applicable to Class A, B, C, D, G, and H amplifiers. At the same time, it simplifies bridge circuits, as well as push-pull configurations.

Fig. 1-15 shows the basic concept of complementary symmetry. Note that the bases of the power transistors are connected together. This might not initially appear conducive to push-pull operation, but a little consideration will reveal that each transistor is made conductive by its own half cycle of an ac input signal. Also Class B operation automatically ensues. Actually, in order to eliminate the dead zone (where neither transistor is conductive) and to reduce crossover distortion, both transistors are generally provided with a slight amount of forward bias.

An actual audio-amplifier with complementary-symmetry output is shown in Fig. 1-16. The functions of the important elements involved in the operation of the circuit are as follows: Transistors Q4 and Q5 are

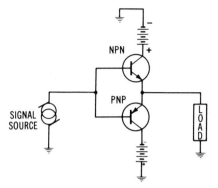

Fig. 1-15. Basic complementary-symmetry amplifier.

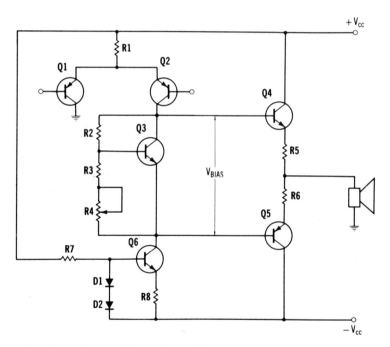

Fig. 1-16. Simplified audio amplifier with complementary-symmetry output stage.

a complementary-symmetry pair operating as a Class AB output stage. (Operation is essentially Class B, with just enough forward bias practically to eliminate crossover distortion.) The fact that the bases of these transistors are not directly connected together stems from their differing bias requirements. The actual application of the signal is such that both bases are essentially driven in parallel. This simple system has but one driver, the differential-voltage-gain stage made up of transistors Q1 and Q2. Providing that an input signal with sufficient amplitude is available, such a signal could be applied to the base of Q1. A resistive network could then be inserted between the junction of R5 and R6 and the base of Q2 in order to provide negative feedback. Transistor Q3 supplies the forward biases to the complementary-symmetry output stage. This could be a ticklish task, for it is known that the base-emitter voltage of each of the output transistors will decrease at the rate of 2 millivolts per degree Celsius. Besides supplying these transistors with different biases— one being npn, and the other pnp—some automatic tracking arrangement must be incorporated to prevent the output idling current from changing with temperature. Transistor Q3 is able to provide both of these biasing requirements for two reasons. First, it is fed from a constant-current stage, Q6, and second, it is physically mounted on the same heat sink supporting output transistors Q4 and Q5. The constant-current source makes V_{BE} of Q3 dependent on *temperature* only. The heat-sink mounting technique constitutes thermal feedback, which causes Q3 to provide compensatory biases in response to a temperature change. The V_{BE} of Q3 can be thought of as the "reference" voltage in such a thermal–electrical feedback arrangement. Variable resistance R4 adjusts the nominal biases applied to the output transistors.

Suppose that increased temperature tends to cause the quiescent, or idling, current of the output stage to increase. The same temperature increase causes V_{BE} of Q3 to decrease, thereby decreasing the biases applied to the output stage and nearly restoring the idling current to its original value. Good tracking can occur over a wide temperature range because the physical mechanisms governing output current and applied bias from Q3 are essentially similar—both are predicated on V_{BE} responses to temperature.

Actually constant-current source Q6 serves *another* important function. It acts as a high-impedance dynamic load for transistor Q2 in the input stage. This enables the input stage to develop a much higher voltage gain than is readily accomplished with a collector load resistance.

Not only do complementary-symmetry power amplifiers make it easy to dispense with transformers, but it also becomes unnecessary to use a large electrolytic coupling capacitor in the speaker or load circuit when dual power supplies are used.

THE QUASI-COMPLEMENTARY-SYMMETRY AMPLIFIER

Basic quasi-complementary-symmetry amplifiers are shown in Fig. 1-17. One of the advantages of this type of circuitry over the "pure" complementary-symmetry arrangement is that a pair of identical power transistors is used in the output stage. The reason that the quasi circuit came into prominence was because the earlier germanium pnp transistors did not have any germanium npn counterparts. Also the later silicon npn transistors usually did not have a silicon pnp counterpart. Indeed it has only been relatively recently that silicon npn-pnp matched pairs have been readily available on the market. Fewer silicon pnp transistors are still available for higher power ratings.

Fortunately the main features of complementary-symmetry circuitry are attainable from arrangements that make one output transistor behave as though it were of opposite polarity. The basic idea is to use a conventional Darlington connection from a driver on one side of the push-pull circuit and a complementary-symmetry Darlington on the other

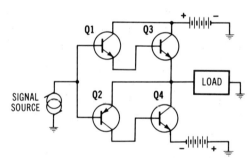

(A) Circuit using npn power output transistors.

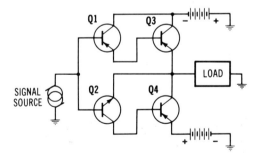

(B) Circuit using pnp power output transistors.

Fig. 1-17. Basic circuits of quasi-complementary-symmetry amplifiers.

side. Thus, for the circuits in Fig. 1-17 the upper portion of the circuits (involving transistors Q1 and Q3) comprise *conventional* Darlington combinations of driver and output transistors. Conversely the lower portion of the circuits (involving transistors Q2 and Q4) utilizes the *complementary-symmetry* type of Darlington connection. Insofar as the signal source, power transistor Q4 of the circuit in Fig. 1-17A behaves as if it were a pnp rather than an npn type. Thus, the output phase relationship between the two halves of the circuit complies with the needs of push-pull operation. In the circuit in Fig. 1-17B power transistor Q4 behaves as if it were actually an npn type.

One might question the degree of balance prevailing when one transistor of the push-pull output pair delivers load current from its emitter, and the other delivers load current from its collector. Surprisingly this is not a major source of imbalance in audio amplifiers, primarily because of the low load impedance presented by speakers. Under such a condition, common-emitter and common-collector circuits operate in a similar fashion. Even so, some manufacturers produce specially designed power transistors intended for optimum performance in quasi-complementary-symmetry amplifiers.

The circuits of Fig. 1-17 are simplified and do not include biasing provisions. As depicted, they would produce crossover distortion in audio applications but would be entirely suitable for certain other uses, such as servo amplifiers. If the circuits shown were used, the relative severity of crossover distortion would be much less if 50- or 60-volt transistors were used than if the amplifier were designed for 12- or 15-volt operation.

CLASS C AMPLIFIERS

The Class C amplifier carries the operating concept of Class B amplification even farther. In a Class C amplifier output current flows for only a small portion (120 degrees or less) of an input sine wave. Therefore, two devices operated in Class C mode in a push-pull arrangement cannot restore the missing part of the input signal, and linear amplification cannot be accomplished. This rules out the use of such an amplifier for reproducing audio signals. The Class C amplifier finds its principle application in boosting the power level of radio-frequency signals in transmitters. A *resonant circuit* in the output stage provides the means whereby a sustained radio-frequency sine wave is produced for the antenna, or for another Class C stage. In essence the resonant circuit is "shock excited" into oscillation. Despite its name the Class C amplifier behaves more like a *switch* than an amplifier. If, for example, it were possible to drive an electromechanical relay at a radio-frequency rate, the same type of operation would ensue if the contacts of the relay were associated with a dc power source and a resonant circuit. Think of the Class C amplifier as a source of current pulses. This being the case, it is only natural for the resonant circuit to oscillate. Although the oscillations so produced are *damped,* they are not perceived to be so if the Q of the resonant circuit is high enough. The *flywheel* effect of such a *tank* circuit then produces an essentially constant radio-frequency output.

Sometimes it is desirable to impart power amplification to a *modulated* radio-frequency signal. If this were attempted with a Class C amplifier, either a single-ended or a push-pull configuration, the audio modulation would not be reproduced in the output of the resonant tank circuit. Such amplification can, however, be accomplished with amplifier circuits that resemble Class C stages but are biased to operate in the Class B mode. Such an amplifier is appropriately termed a *linear Class B stage.* It increases the radio-frequency (rf) power level by shock exciting the resonant output circuit with current pulses. (Of course, the current pulses are now 180 degrees in "pulse width.") Its operation resembles the switching action of the Class C amplifier. However, the radio-frequency oscillations excited in the output tank circuit will *follow* the amplitude undulations of the modulated input signal. Either single-ended or push-pull configurations work well in this application.

Class A amplifiers associated with a resonant output tank circuit can also boost the power level of a modulated rf signal. The chief drawback of Class A rf power amplification is its relatively low operating efficiency. On the other hand the generation of radio-frequency harmonics is much lower than with either Class B or Class C. Fig. 1-18 illustrates the basic operational differences in Class A, B, and C amplifiers.

CLASS D AMPLIFIERS

The development of solid-state power devices has created new interest in techniques that have their roots in the past but could not be readily exploited during the long reign of the vacuum tube. A Class D amplifier

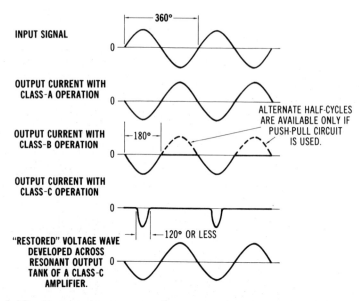

INPUT SIGNAL

OUTPUT CURRENT WITH
CLASS-A OPERATION

OUTPUT CURRENT WITH
CLASS-B OPERATION

OUTPUT CURRENT WITH
CLASS-C OPERATION

"RESTORED" VOLTAGE WAVE
DEVELOPED ACROSS
RESONANT OUTPUT
TANK OF A CLASS-C
AMPLIFIER.

ALTERNATE HALF-CYCLES
ARE AVAILABLE ONLY IF
PUSH-PULL CIRCUIT
IS USED.

Fig. 1-18. Input and output waveforms associated with Class A, B, and C operation of power devices.

exemplifies such a situation. Like a Class C amplifier, the Class D amplifier involves pulsed operation and requires an LC output circuit. However, that is where the resemblance ends. The Class D operated device produces a pulse-width-modulated waveform that bears the information extracted from an input signal. The LC circuit is a low-pass filter with sufficient damping or loading to *prevent* the type of ringing characteristic of the resonant tank of a Class C amplifier. As a consequence, the output of the low-pass filter is a power-boosted version of the input signal. Viewed as a black box, the Class D amplifier functions as a linear amplifier. That is, one need not know about the strange things occurring between the input and the ultimate output of this amplifying system. In order to perform well, the low-pass output filter should do a good job of removing switching-frequency components. This is facilitated by making the switching rate much higher than the highest frequency to be amplified. Figs. 1-19 and 1-20 provide additional insights into the operation of Class D amplifiers.

Sometimes the nomenclature *Class D* is applied to *other* switching-type circuits. The operation of an inverter, particularly a driven inverter,

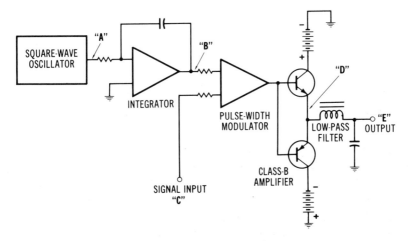

Fig. 1-19. The basic Class-D or "Two-state" amplifier system.

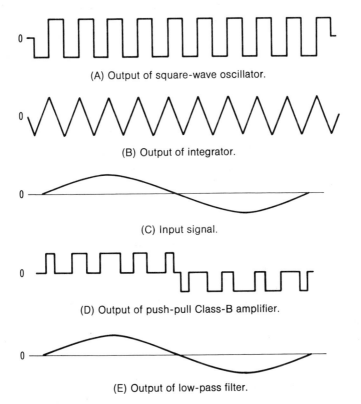

(A) Output of square-wave oscillator.

(B) Output of integrator.

(C) Input signal.

(D) Output of push-pull Class-B amplifier.

(E) Output of low-pass filter.

Fig. 1-20. Waveforms from Class D amplifier system in Fig. 1-19.

41

has been referred to as Class D operation in technical literature. The switching device in the dc switching-type regulator has also been referred to as Class D operation. Here the resemblance to the amplifier just described is somewhat closer because the switching device is often keyed by a pulse-width-modulated waveform. Moreover, the output low-pass filter performs a similar function to the low-pass filter in the Class D amplifier, which is the removal of frequencies associated with the switching rate. In fact the commonly encountered 20-kHz switching regulator could be readily converted into an audio power amplifier (not hi-fi). This could be accomplished by superimposing the audio input signal on the reference voltage. This possibility has been cited in order to emphasize the importance of this relatively recent technique for processing power with a solid-state device. The desirable feature of switching-type operation is that the solid-state device is either on or off. When it is off, no output current flows, so there is no power dissipation. And when the device is on, only a minimal voltage drop exists between the anode and cathode. Therefore, the power dissipation is *low*. If, furthermore, the rise and fall times of the switching transition are small compared to the width of the narrowest pulse being handled, the overall internal dissipation of the device will be *low* and its operating efficiency will be *high*.

THE CLASS F AMPLIFIER

The Class F amplifier is capable of higher efficiency in the boosting of radio-frequency power than are the Class B and Class C amplifiers traditionally used for this purpose. This is particularly true when a power MOSFET is utilized in the circuit because this device consumes negligible input power. The maximum operating efficiency of the Class B amplifier is 78.5 percent, and that of the Class C amplifier is somewhat higher. However, ordinary Class C amplifiers do not simultaneously develop maximum efficiency and maximum power output. In contrast the ideal efficiency of the Class F power amplifier is 100 percent. Inasmuch as Class F operation is closely related to that of Class B amplification, it will be more relevant to compare the Class B and Class F operational modes.

Fig. 1-21 is a representative circuit of a Class F power amplifier. Because the MOSFET is an enhancement-type device, the biasing provision is automatically suitable for Class B operation wherein output

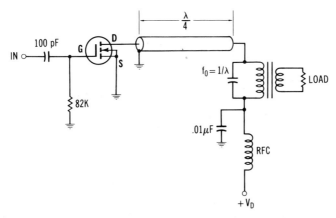

Fig. 1-21. Basic Class F radio-frequency power amplifier.

(drain) current flows for 180 degrees of the sine-wave input (gate) signal. Indeed Class B operation would occur, except for the effect of the quarter-wavelength transmission line inserted in the drain circuit. The effect of this unusual feature can be discerned from inspection of the waveforms in Fig. 1-22.

In Fig. 1-22 the important waveforms are those of drain voltage and drain current. The dashed sine wave shown in the drain-voltage waveform depicts the situation that would prevail without the presence of the quarter-wavelength transmission line. This would then be Class B operation.

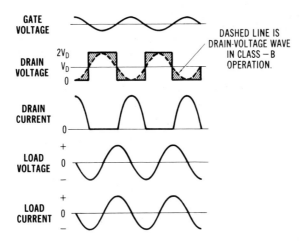

Fig. 1-22. Waveforms in the Class F radio-frequency power amplifier.

In Class B operation there are times when we have both drain voltage and drain current. The simultaneous occurrence of both voltage and current results in power dissipation and is the reason for the theoretical limit of efficiency. In Class F operation the drain voltage is a square wave. As can be seen in Fig. 1-22, the drain voltage is zero when there is drain current. Also the drain current is zero when drain voltage exists. This is a "pure" switching action; hence there is no internal dissipation and ideal operating efficiency is 100 percent. Of course, this ideal condition does not exist when the MOSFET is operating in its linear, or power-dissipating, mode.

It appears that the function of the quarter-wavelength line, in simplistic terms, is to convert the drain voltage from a sine wave to a square wave. This indeed is true. Inasmuch as the quarter-wavelength line is neither an active nor a nonlinear device, it is instructive to deduce the nature of its waveshaping property. In Class B (and Class C) operation, it is known that the resonant output circuit restores half-sine waves of a pulse train to a continuous sinusoidal waveform. How it accomplishes this can be explained in several ways. The explanation bearing greatest relevance to the topic under discussion is the selection and rejection of harmonic energy. Specifically the parallel-resonant load circuit rejects all harmonics of the fundamental frequency. It does this by acting as a short circuit to these harmonics. At the same time it *supports* the fundamental frequency, f_o. It does this by appearing as a high impedance to the fundamental frequency. (From the Fourier theorem of wave analysis, all nonsinusoidal waves are fabricated from sine-wave "building blocks." These sine waves are a fundamental frequency, f_o, and a number of harmonic frequencies, $2f_o$, $3f_o$, $4f_o$, $5f_o$, etc. The number, magnitude, and relative phases of the harmonics determine the shape of the wave. Only the sine wave itself is free of harmonics.)

Let us investigate what happens when the harmonic-discrimination properties of a quarter-wavelength transmission line are introduced into the circuit. For even harmonics, that is, $2f_o$, $4f_o$, $6f_o$, etc., the quarter-wavelength line behaves as a half-wavelength line. Such an element simply *reproduces* the load situation presented to the MOSFET. Inasmuch as the parallel-resonant load supports f_o and short-circuits *all* harmonics, it follows that no change is thus far indicated. For odd harmonics, that is f_o, $3f_o$, $5f_o$, $7f_o$, etc., the quarter-wavelength line displays the significant property that it *inverts* the impedance of the load, thereby allowing the odd harmonics of f_o to appear at the drain. Although f_o

itself is, analytically speaking, an odd harmonic, the quarter-wavelength line does *not* reflect f_o at the load back to the drain. The differential treatment accorded to f_o stems from the fact that the load appears as a high impedance at f_o, while it appears as a short circuit to the *other* odd harmonics.

The *objective* is to impress a large number of odd harmonics of f_o at the drain, for such a combination, together with f_o, will synthesize the square-wave drain voltage needed for Class F operation. So far we have accounted for the accumulation of these odd harmonics, but it has been shown that the quarter-wavelength line will not return f_o itself from the load to the drain. Fortunately this is not necessary because f_o already is present at the drain as represented by the dashed sine wave in Fig. 1-22. Accordingly the odd harmonics returned by the quarter-wavelength line combine with the sine wave already present at the drain and produce the requisite square wave.

The shaded area between the two drain voltage waves in Fig. 1-22 is where dissipation takes place in Class B operation. By causing the drain voltage to be a square wave, such dissipation is *eliminated*.

It might be supposed that the Class F amplifier is applicable only to the uhf region, where a quarter-wavelength transmission line assumes practical linear dimensions. However, this interesting circuit can be implemented at lower frequencies within the amateur radio spectrum by using coiled coaxial cable. Coiling of coaxial cable will not appreciably affect its transmission line properties as long as the winding diameter is much greater than the diameter of the cable itself. Microdot miniature coaxial cable can be used to reduce packaging size further.

The Class F power amplifier can be drain modulated. The parallel-resonant load circuit can be replaced by a pi network or by a broadband transformer. With the Siliconix VMP-1 MOSFET, 11 or 12 watts of output should prove feasible. With the newer VN84GA 12.5-ampere, 80-volt MOSFET, several times this power level should be readily forthcoming. In any event the main deterrent to ideal operating conditions is the on resistance of the MOSFET. Unfortunately this is not zero, as postulated in our concept of ideal switching action. Therefore, the realizable efficiency will tend to be in the area of 80 percent, at best. On the other hand there is negligible loss associated with switching times. The Class F amplifier, together with MOSFET power for radio-frequency amplification, is relatively new and is fertile ground for the experimentally inclined.

CLASS G AMPLIFIER

Even the Class B amplifier, or the Class AB amplifier, cannot develop high efficiency when handling audio waveforms. Because of the high peak-to-average ratio of audio signals, and because low amplitudes prevail *most* of the time, the amplifier operates most of the time in its low-power region where efficiency is necessarily low. But the amplifier must nonetheless be designed to accommodate the peaks. One naturally wonders whether something can be done about this situation.

If it were merely required to actuate an electric motor, it would be permissible to *clip* the peak excursions. Inasmuch as peaks are invested with relatively little energy, their absence would not be noticeable. Surprisingly the same technique *can* be imparted to speech waves. Although the characteristics of the reproduction are altered, considerable clipping is allowable before *intelligibility* suffers. However, for high-fidelity reproduction of speech, and especially of *music,* peak-clipping is intolerable. Indeed the ability to handle even occasional peaks explains the reason for 200-watt amplifiers, which provide adequate volume in an average room while delivering less than a watt of output power.

The solution to this problem is found in the relatively new Class G amplifier. With this amplifying system a push-pull Class B output stage has its dc supply voltage automatically switched in such a way that good operating efficiency prevails for both low- and high-amplitude input signals. More specifically the output stage operates with low dc collector voltage when amplifying small signals and with high collector voltage when amplifying large signals. Inasmuch as the optimum efficiency of Class B amplifiers is attainable *only* when the collector voltage swing approaches the dc collector voltage level, the average efficiency of Class B operation when processing audio signals falls considerably short of its potential. But when the Class B stage is incorporated into the Class G system, the average efficiency is much higher. The dividends reaped by this technique include various combinations of higher peak-power capability, reduced thermal problems, lowered manufacturing costs, and increased rms output power. The predominate feature depends to a large extent on the ratio of the high dc supply voltage to the low dc supply voltage. In practice a reasonable blend of these features prevails when the low-voltage supplies provide about one-half the dc voltage of their high-voltage counterparts.

In the simplified Class G amplifier shown in Fig. 1-23 the Class B

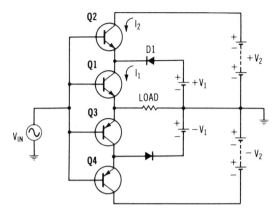

Fig. 1-23. Simplified circuit of the Class G amplifier.

output stage is the complementary pair of transistors, Q1 and Q3. Transistors Q2 and Q4 are *not* amplifying transistors but serve as *switches* to transfer operation from supply voltage V1 to supply voltage V2 at the approximate time when the amplitude of the input signal exceeds the dc voltage level of the V1 power supplies. To understand how this occurs, consider only the operation of the upper portion of the circuit, which includes transistors Q1 and Q2, diode D1, the positive V1 and V2 supplies, and the load. Initially let the amplitude of the input signal be *less* than V1. Under this operating condition the load receives its current from supply V1 through diode D1 and transistor Q1. At the same time transistor Q2 is reverse biased and is therefore inactive. Next suppose that the input signal increases in amplitude to the extent that it exceeds the dc level of supply voltage V1 (but it does not exceed the voltage level of supply V2). Under this operating condition transistor Q2 becomes forward biased and turns on. Current from supply V2 flows through the collector-emitter circuit of Q2, through transistor Q1, and then through the load back to supply V2. Thus, transistor Q2 has acted as an automatic switch to raise the dc collector voltage of Q1 when needed to accommodate higher power levels. As soon as transistor Q2 turns on, diode D1 becomes reverse biased, thereby *preventing* participation of supply V1. In its reverse-biased state, diode D1 *also* prevents current from supply V2 from establishing a path back through the lower-voltage supply, V1. Fig. 1-24 illustrates the gain in efficiency over conventional Class B operation.

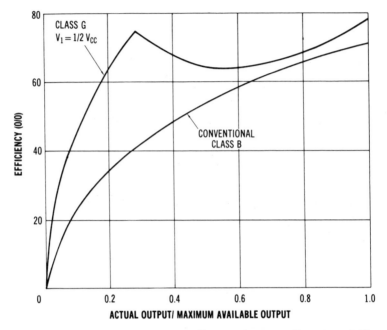

Fig. 1-24. Comparison in operating efficiency between Class B and Class G amplifiers.

Actually, when this technique is used in hi-fi audio applications, certain minor modifications are made. For, as one might suspect, the switching transition may not be as smooth as would be desired. In order to prevent, or reduce greatly, the approximately 0.6-volt "deadzone" due to the emitter-base contact potential of switching transistor Q2, a forward-conducting diode can be placed in the base lead of Class B transistor Q1. Because such diodes also provide protection against excessive base-emitter reverse bias, they are often found in the base lead of *all* of the transistors. Because of charge-storage effects in the switching of transistor Q2 (and its counterpart, transistor Q4), a small *inductance* may be inserted in series with the load. However, these modifications do not materially change the basic operation as described. Also the lower half of the circuit operates in the same manner as the upper half when the polarity of the input signal reverses. However, the polarities of Q3 and Q4 are reversed with respect to Q1 and Q2.

THE CLASS H AMPLIFIER

The Class H amplifier is an even more sophisticated solution to the tendency of conventional amplifiers to develop low efficiency with audio waveforms. Like the Class G amplifier, the objective of the Class H amplifier is to prevent application of more dc voltage to the output stage than is necessary to handle the instantaneous signal amplitude. For it is this *excess* of input dc power that causes unnecessary power dissipation—even the Class B or AB output stage develops its high efficiency *only* when the signal amplitude closely approaches the level of the dc voltage from the power supply. In many servoamplifier applications, continuous high efficiency can be provided by a Class B output stage because the sine-wave signal is always at its maximum amplitude. As pointed out, audio waveforms are far different from this situation because of their high ratio of peak-to-average amplitude. And because they operate for long periods at low amplitudes, only about 20-percent efficiency can be expected from a Class B or AB output stage rather than a reasonable approach to the theoretical maximum efficiency of about 78 percent.

It is evident, however, that the disparity between ideal and actual efficiency could be greatly alleviated by a technique that would provide high dc voltage to the output stage *only to the extent needed.* To be practical, such a technique would actually have to start increasing the supply voltage *before* the output amplifier actually required it. This indeed is just what the Class H amplifier accomplishes. A significant advantage over the Class G amplifier is that the output amplifier itself is not called on to make any switching transitions. Rather all adaptations to signal amplitude are made by power supplies that are actuated by the output signal amplitude. Therefore, the power-output stage is a relatively simple push-pull Class AB circuit with conventional feedback.

In the simplified circuit of Fig. 1-25 voltage comparators with high gains and slew rates turn on series-pass control transistors when output signals approach a set level. *Anticipation* of high-amplitude signal swings occurs because of differentiator networks in the comparator sensing leads. The transfer function of an RC differentiator network is such that its output voltage is proportional to the *rate of rise* of the input signal. It appears plausible that the sensing could be accomplished in a preamplifier, rather than in the output, stage. Then the propagation delay in the preamplifier and in the driver stages could be advantageously used

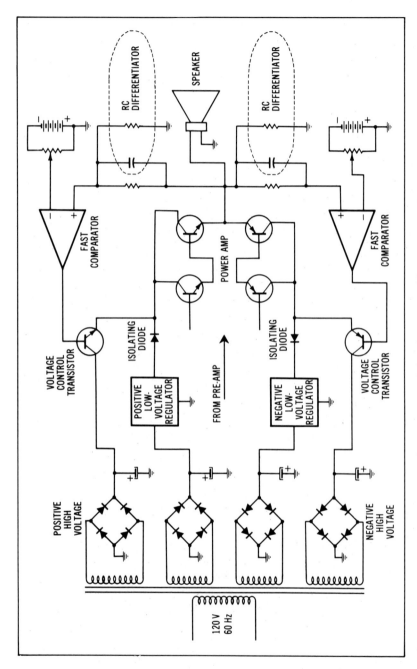

Fig. 1-25. Simplified circuit depicting basic concept of the Class H amplifier.

to help achieve the anticipatory action. The anticipatory action and supply voltage adaptation in the Class H amplifier are illustrated in Fig. 1-26. The low voltages are about 65 percent of the high-voltage levels.

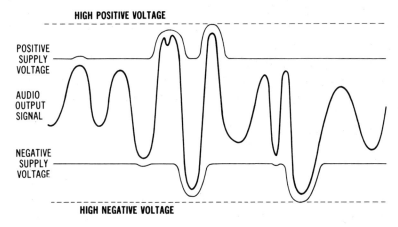

Fig. 1-26. Anticipation and adaptation of power supply voltages in the Class H amplifier.

2

Theories Pertinent to Solid-State Power Devices

Working with *power* electronics is much like working with ordinary electronics except that there is a different *emphasis* on device characteristics. For example, such concepts as frequency response, current gain, and voltage polarity retain their basic meanings. On the other hand one must focus more attention on thermal impedance, charge storage, SOA (safe operating area), thermal runaway, and transients. And certain "simple" functions such as rectification or switching are found to exhibit complexities that, though of trivial importance in ordinary electronics, require careful attention in many power applications. For example, one might naively suppose that a certain low-power inverter circuit using SCRs can be simply scaled up to produce an inverter with, say, 10 times the power-handling capability. But after increasing the ratings of all components including the SCRs by appropriate factors, it may be found that the high-power version of the circuit refuses to operate. Such peculiarities will be discussed in later paragraphs. To start with, let us investigate certain power aspects of the simplest of solid-state elements, the pn diode.

THE PHYSICS OF SEMICONDUCTOR BEHAVIOR

Our primary concern in this book is with *circuit applications of solid-state power devices*. This being the case, it is convenient to consider a device as a circuit component with input and output terminals. That is, we prefer not to become excessively involved with events transpiring *within* the device—the fact that such events occur gives rise to the parameters and characteristics of the device. Such a "black box" approach is quite effective in practice. However, much more insight into circuit action results from awareness of the basic nature of semiconductor behavior. For then we find ourselves in a *better* position to incorporate solid-state devices into appropriate circuits. Instead of merely assuming that a diode rectifies because it passes current much more readily in one direction than the other, it will better serve our purposes to see *why* it exhibits this unilateral feature. It will then prove much easier to cope with certain behaviors of the solid-state devices that require special consideration in power electronics. Our involvement with semiconductor physics will be relevant but minimal. Those desiring a more detailed study are referred to the specialized texts providing comprehensive treatment of this subject.

ELECTRICAL CONDUCTION IN METALS

The use of metallic wires to conduct electrical current generally does not produce a second thought when we connect the various components in a circuit. Yet such conductors are also in essence components. Their conductive ability is as much a solid-state phenomenon as is transistor action in a semiconductor. Indeed it may appear in retrospect that familiarity with the conduction of electrons in metals and the opposite behavior in insulators should have inspired earlier development of solid-state devices. This, of course, is an easy statement to make but a tantalizing one too. For in a qualitative way it is true that diodes, transistors, and other semiconductor devices depend on material that is *neither* a good conductor nor a good insulator but rather somewhere in between—thus the prefix *semi*. But as we shall see, merely qualifying for classification as a poor conductor, or as a poor insulator, does not suffice for a material to be made into a diode or a transistor.

A metal such as copper is an elemental substance composed of copper atoms. The formation of the gross substance, copper, involves relatively tight packing of these atoms. Indeed the atoms are thought to be so close together that their outer electronic orbits overlap as shown in Fig. 2-1. Such electrons are not uniquely bound to any one particular

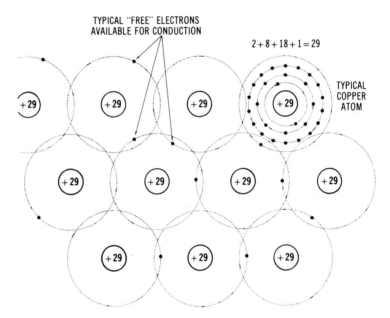

TYPICAL "FREE" ELECTRONS
AVAILABLE FOR CONDUCTION

2 + 8 + 18 + 1 = 29

TYPICAL
COPPER
ATOM

Fig. 2-1. Outer orbits of copper atoms share electrons.

atom; rather these electrons exist in the material as a "fog" surrounding the atoms. Assuming that this description is substantially valid, it is next postulated that random motions are imparted to these electrons by the action of thermal energy. Interestingly this erratic displacement of free electrons produces pulses of electric current, which may be audibly or visibly detected with the aid of a high-gain amplifier. This so-called "noise" actually embraces a wide frequency spectrum. However, no sustained flow of electricity takes place in such a copper conductor.

If we subject the free electrons to the pressure of an electric field by connecting a small battery across the ends of a copper wire, an orderly procession of electrons is *superimposed* on the thermally induced random motion. Electrons are then supplied by the negative terminal of the battery and collected at the positive terminal. Within the wire, electrons are impelled to move from one atom to another. Other than thermally induced vibration, the bulk of the atom is *not* free to move. Also it should not be thought that any particular electron injected into the negative end of the wire makes a speedy transition to the positive end. Rather what we call a *flow* of electrons is more precisely a relaying process wherein the impetus received at the negative end of the wire is communicated with great speed to an electron at the positive end of the wire. The latter electron is ejected from the wire and collected at the positive terminal of the battery.

Thus, while it is true that electricity traverses such a conductor with a speed close to that of light, the motion of *individual* electrons from one end of the wire to the other is more in the nature of a slow drift. The basic mechanism involved may be demonstrated by arranging a row of marbles on a table. An impact conveyed to an end marble is quickly transmitted to the marble at the opposite end, which is expelled from the row. The impacting "shooter" marble does not have to traverse the length of the row in order to eject the marble at the opposite end. Accompanying the ejection of the final marble was a "drift" of all marbles in the same direction toward which the final marble was ejected.

In our simplified model of conduction in metals, it is well to reflect that a *charge carrier* (the electron) is required. Although premature in the development of pn-junction theory, it is interesting to speculate the result of exerting control over the *availability* of the free carriers. The result would be some kind of a "solid-state" relay or amplifying device. Unfortunately no practical way of doing this with an ordinary metallic conductor has evolved.

ELECTRICAL CONDUCTION IN LIQUIDS

Having accounted for electrical conduction in metals, we will next consider liquids and gases, which will bring us somewhat closer to the conduction phenomenon in semiconductors. Let us see why. A liquid does not conduct in the manner of a metal because the atoms are not so densely compacted as in metals. Indeed many liquids are excellent insulators for they are devoid of mobile charge carriers. The electron "fog" described for metals is not present in these liquids because orbital electrons in the atoms remain in captivity of the electrostatic forces of the atomic nuclei. Oils and alcohols generally belong to this category. Also pure water tends to be an excellent insulator. Our usual conception of water identifies it with conductive properties. However, this is because we seldom encounter water with a high degree of purity. Because of the high dialectric constant of water, the electrostatic binding forces between the atoms of *other* molecules are weakened. The liberated atoms then become ions—that is, they bear electric charges. Such ions enable the solution to conduct electricity. Conduction in a solution of sodium chloride is shown in Fig. 2-2. If we are not interested in the conductive properties of the solution, or electrolyte, we can sum up the situation by classifying water as a good *solvent*. Fig. 2-3 illustrates the formation of crystalline sodium chloride. This atomic array is disassociated by dissolving the salt crystals in the water.

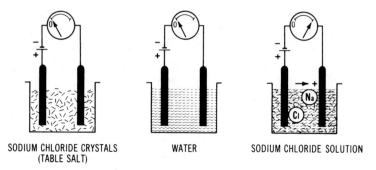

SODIUM CHLORIDE CRYSTALS   WATER   SODIUM CHLORIDE SOLUTION
(TABLE SALT)

Fig. 2-2. Although neither salt nor water alone provides charge carriers, their solution does.

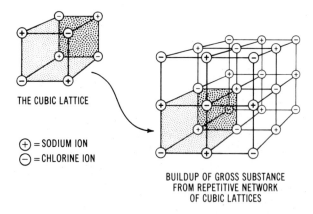

THE CUBIC LATTICE

⊕ = SODIUM ION
⊖ = CHLORINE ION

BUILDUP OF GROSS SUBSTANCE
FROM REPETITIVE NETWORK
OF CUBIC LATTICES

Fig. 2-3. Formation of crystalline sodium chloride from repetitive arrays of ions.

A remarkable thing about many chemical reactions is that the participant atoms unite because of the tendency to have eight electrons in their outermost orbit. Somehow the atoms then assume a more *stable* existence. During the course of the reaction electrons are given up, taken on, or shared until this stable condition is obtained. We then say that the reaction has gone to completion. An example touched on was that of common table salt or sodium chloride. The sodium and chlorine atoms become *ions* in this reaction, but their final status is mutually beneficial because both find themselves with *eight* electrons in their outer orbit as shown in Fig. 2-4. We should not be surprised to find that the *inert*-gas atoms, neon, argon, krypton, and xenon *already* have eight electrons in their outer orbits (Fig. 2-5). The reluctance of these gases to participate in chemical reactions is due to the fact that such participation would tend to *deprive* them of their stability. Let us investigate the involvement of the magic number *eight* in the semiconductor material silicon.

An interesting aspect of electrical conduction in a liquid such as the sodium-chloride solution is that electrical conduction involves the movements of not just one charge carrier as in metals but of *two*. The two carriers, the sodium ions and chlorine ions, bear *opposite* charge polarities and therefore move in *opposite* directions. Their contributions to the electric current are *additive*. This is an important idea to keep in mind, for it is relevant to the electrical phenomena in semiconductor material to be discussed shortly. It is interesting to contemplate that whereas electrical current in the external wires connected to an electrolytic cell is derived from the movement of *electrons,* the internal current of the cell is *not* the consequence of electrons being transferred from one electrode to the other. Although the charges on the ions are due to a surplus or deficit of electrons, the *entire ions* move to the two

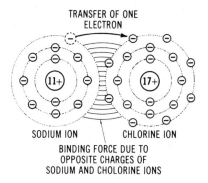

TRANSFER OF ONE
ELECTRON

SODIUM ION   CHLORINE ION
BINDING FORCE DUE TO
OPPOSITE CHARGES OF
SODIUM AND CHLORINE IONS

Fig. 2-4. Electronic phenomena in the chemical reaction producing sodium chloride.

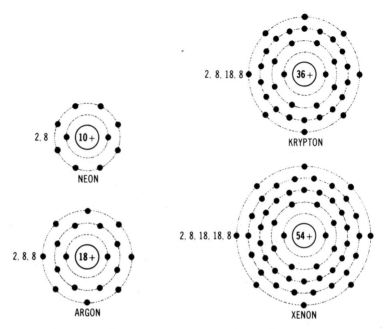

Fig. 2-5. The atoms of these four inert gases exemplify the stabilizing effect of eight electrons in their outer orbits.

electrodes, where they are "collected." In this process the ions again become atoms, and thus we should expect to find a concentration of metallic sodium at the negative electrode and the evolution of chlorine gas at the positive electrode. (Actually the sodium metal, being extremely reactive, quickly gets involved with other reactions in the solution. For example, some recombines with free chlorine and reforms the original salt or sodium chloride.) The concept of recombination should also be kept in mind, for a somewhat analagous occurrence is important in the operation of semiconductors.

ELECTRICAL CONDUCTION IN GASES

The physical characteristics of a gas are attributed to relatively large separations between the constituent atoms. Consider, for example, neon gas. Neon is classified as an *inert* element, implying a weak affinity for other elements. This is so because the neon atom *already* has eight electrons in its outer orbit. It need not take on, give up, or share electrons with other atoms to improve its own stability. One might then suppose that neon gas is an excellent insulator. This is true; but paradoxically the extensive use of neon in the electrical industry generally makes use of its *conductive* properties.

Neon gas, although a "natural" insulator for the aforementioned reasons, can be made to behave as a conductor by supplying it with a certain minimum energy. This added energy frees one or more of the outer-orbit electrons from the attraction of the nucleus. When this occurs, the neon atom is *ionized* and can function as a charge carrier. The freed electron is also available as a mobile charge carrier. Moreover, the ion and the freed electron may strike other atoms, thereby spreading the ionized state throughout the gas. Neon ions are ultimately collected at the negative electrode, whereas free electrons go to the positive electrode. However, both of their motions are *additive* with respect to the total electric current in the ionized gas. Electrical conduction in neon gas is depicted in Fig. 2-6.

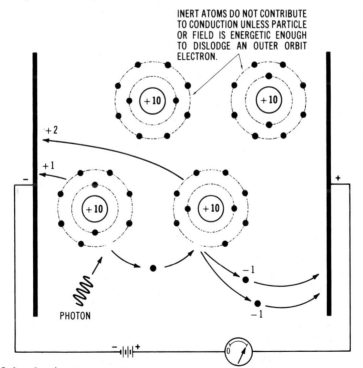

Fig. 2-6. Conduction in neon gas.

The added energy needed to ionize neon gas can take several forms. For example, it is readily supplied in the form of an electric field. Raising the voltage across two electrodes situated in a neon-filled container ultimately attains a level sufficient to break down, or *ionize,* the gas. In many practical devices this occurs between 65 and 85 volts. Ionization is similarly produced by cosmic-ray particles, X-rays, and radio-frequency fields. Thermal energy can be supplied to either produce ionization or to facilitate its occurrence by other forms of energy. Electron beams, too, readily impart ionization energy. The common denominator in all cases is the *liberation* of one or more of the outer-orbit electrons in the neon atom. The study of electrical discharge in gases is complex and we have merely glossed over the basic principles. However, these highlights suffice to provide useful analogies with electrical phenomena in semiconductors.

THE ELECTRICAL NATURE OF INTRINSIC SILICON

Having discussed the several ways in which electrical conductivity can occur, or be brought about, we find ourselves in a more favorable position to deal with semiconductor material. Intrinsic, or *pure,* silicon at ordinary temperatures is a solid crystalline substance made up of silicon atoms. The individual silicon atom has four electrons in its outer orbit. This unique arrangement allows these atoms to unite with their *own kind* in the manner illustrated in Fig. 2-7. In such a formation, known as *covalent bonding,* each atom has attained the stable condition in which its outer orbit comprises *eight* rather than four electrons. As mentioned, eight outer electrons tend to confer stability to most atoms. Accordingly, pure crystalline silicon should be a good electrical *insulator,* because all electrons are retained in the force fields of the atomic nuclei. This prediction does not completely comply with observed behavior, however. Actually there is a *low* conductivity in silicon at room temperature when photons force electrons in the outer orbit across the forbidden energy band into an unstable orbit. This conductivity increases exponentially with temperature and becomes significant at elevated temperatures. The implication of this is that the covalent bonds are not as strong as the binding forces prevalent in the atomic structure of other insulating materials. Thus, the introduction of relatively little energy suffices to liberate occasional electrons, which are then available to carry the electrical current. This is illustrated in Fig. 2-8.

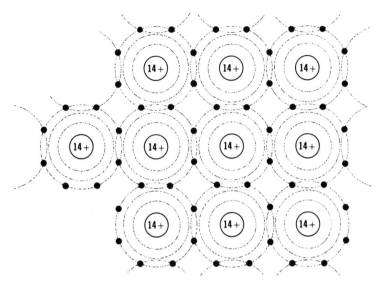

Fig. 2-7. Covalent sharing of electrons in silicon.

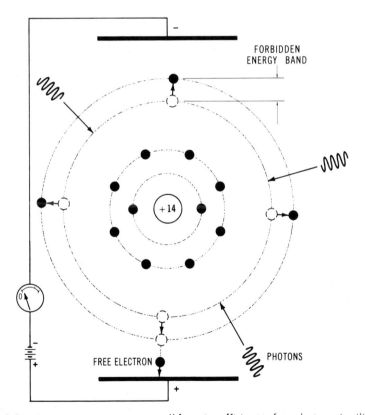

Fig. 2-8. At room temperature a small force is sufficient to free electrons in silicon.

At room temperature, silicon displays some of the aspects of metallic, liquid, and gaseous conduction. See Fig. 2-9. It resembles the behavior in metals in that the ionized atom itself cannot move through the substance. Intrinsic silicon is somewhat suggestive of a "weak" metal in which sparsely populated free carriers are available. The resemblance to liquids is the involvement of *two* charge carriers. One of these is the electron; the other is the vacancy, or "hole," left by a moving electron. The hole moves in the opposite direction to the electron and, despite the strangeness of the concept, should be thought of as a free positive carrier. It contributes to current just as the more tangible positive ion does in liquids.

The analogy to gaseous conduction stems from the fact that energy must be added in order to free outer-orbit electrons from individual atoms. In any event we do not yet have junction diodes or transistors resulting from the slight room-temperature conductivity of intrinsic silicon. Apparently much further progress is needed in order to point the way for practical electronic control devices.

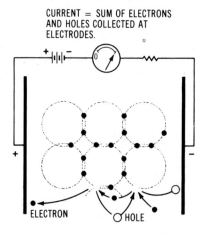

Fig. 2-9.   Conduction in an intrinsic semiconductor material such as pure silicon.

The Effect of Doping a Semiconductor Material

It is known that because of unique processing of intrinsic silicon, solid-state devices made from silicon can exhibit high electrical conductivity and that such conductivity is also subject to *control* by applied electrical signals. Let us explore the means whereby such desirable action can be brought about.

The phosphorus atom is similar to the silicon atom, except that it contains *five,* rather than four, outer-orbit electrons. Suppose that we add "just a pinch" of phosphorus atoms to crystalline silicon. Chemically and physically any changes in the silicon would not, for practical purposes, be discernible. However, a significant *electrical* change would result. Fig. 2-10 shows the incorporation of the phosphorus or so-called *donor* atoms in the covalent bond structure of the silicon material. The situation is similar to that prevailing in "pure" silicon material in that all atoms are stable with *eight* electrons sharing the outer orbit. However, there is this *difference.* The fifth electron in the outer orbit of the phosphoros atom becomes a surplus electron. Therefore, such extra electrons are available in the adulterated or *doped* material as *conduction* electrons. Thus, the electrical conductivity of silicon semiconductor material is greatly increased because of the donor atoms and their extra orbital electrons. (Arsenic and tin, also with five outer electrons, merit consideration as donor atoms. However, the relative merits of each will not be discussed here.)

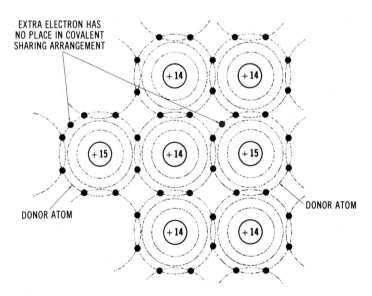

Fig. 2-10.   Donor atoms with five electrons in outer orbit contribute surplus electrons to silicon.

Silicon material doped with donor atoms is known as n-type semiconductor material because conductivity is enhanced via *negative* charge carriers, that is, the electrons.

P-Type Semiconductor Material

The logic of conductivity enhancement in n-type silicon material naturally leads to other speculations regarding processing of semiconductors. What, for example, would be the result of incorporating atoms such as aluminum with *three* outer-orbit electrons into the crystal lattice of silicon? If again, "just a pinch" of aluminum atoms are added, the physical and chemical characteristics of the bulk material (silicon) need not be appreciably affected. However, the electrical situation will then be as depicted in Fig. 2-11. The missing electrons, or *holes,* constitute deficiencies in the covalent bonding structure of the crystalline material. Because of this, there is a strong tendency for such holes to "capture"

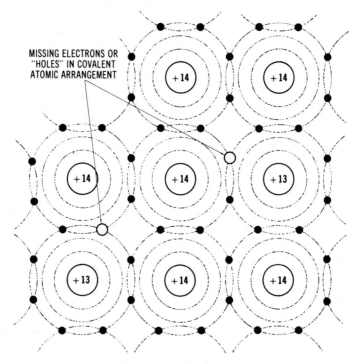

Fig. 2-11. Acceptor atoms with three electrons in outer orbit create "holes" in the covalent bonding structure of silicon.

electrons. The electrons in turn are constantly being liberated from other atomic bonds by thermal energy. The net effect is that the conductivity of the semiconductor material is considerably enhanced by the p-type impurity, or *acceptor* atoms, and that the *main* current carrier is actually the hole. Because this is not easily visualized, it proves more convenient simply to think of the hole as a mobile *positive* charge carrier. The important points to keep in mind are as follows:

1. In *both* n-type and p-type semiconductor material, electrical conductance is greatly enhanced as the consequence of *impurity* or *donor* atoms.
2. Although both negative and positive charge carriers participate in conducting the electrical current in both types of semiconductor, the *predominant* carrier in n-material is the *negative* electron. That is, electrons are *majority* carriers in *n*-material, and holes are minority carriers.
3. Conversely, in p-type material, *holes* are the majority carriers, and electrons are minority carriers.

THE PN JUNCTION

The characteristics of donor-doped and acceptor-doped silicon lead to one of the most important concepts in electronics. Specifically it is the *pn junction,* which makes possible the rectifying diode, the transistor, thyristors, and various other solid-state devices, such as the zener diode, the varactor, and the junction FET.

Let us suppose that two separate regions, one doped with an n-type impurity and the other doped with a p-type impurity, are processed adjacently in silicon material. It is not easy to decide what phenomena should occur at the *junction* of these semiconductor regions. Indeed one could make a good case for mutual annihilation on the one hand or for *no* unusual action on the other hand. Actually a unique formation of atoms does result. What happens is illustrated in four sequential steps in Fig. 2-12.

From this description of the formation of a pn junction one might visualize it as a reservoir of opposite charges paired off side by side. It would be natural to expect that such a condition could be only of a transitory nature, for it would appear that opposite charges must rush

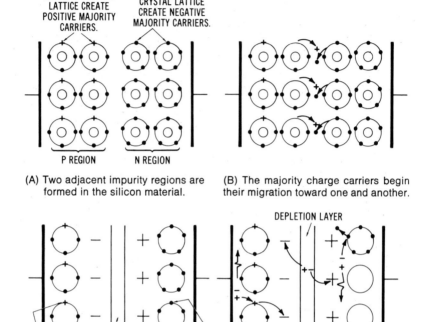

(A) Two adjacent impurity regions are formed in the silicon material.

(B) The majority charge carriers begin their migration toward one and another.

(C) The plus signs indicate fixed positive ions and the minus signs indicate fixed negative ions.

(D) Thermally generated minority carriers maintain parallel arrays of opposite charges as well as a charge-free zone, or depletion layer, between them.

Fig. 2-12.   The formation of a pn junction.

together and recombine. Even if this did not happen, we might suppose that the pn structure would behave as a charged capacitor or as an electronic cell capable of providing current to an external load. By investigating the reasons that there is no charge cancellation, chemical annihilation, or an available current for an external load, we can gain valuable insights into mechanisms basic to the operation of solid-state devices.

Before dealing with the situations depicted in Fig. 2-12, it is well to be aware of the misleading representation of pn diodes so frequently found in technical literature. Thus, in Fig. 2-13A the pictorial symbol lacks an important feature of pn junctions, the depletion layer, as shown in Fig. 2-13B. The depletion layer has the low conduction properties of intrinsic silicon, even though the silicon material has been doped with impurity atoms. A simple way to account for this is to assume that equal numbers of oppositely charged impurity atoms exist in the central region of the pn structure, thereby producing a net cancellation of such charges.

OHMIC CONTACTS   DEPLETION REGION

(A) A commonly encountered drawing.   (B) A more realistic drawing showing the depletion region.

Fig. 2-13.   Representative illustrations of pn diodes.

A more realistic approach involves a dynamic, rather than static, concept of charge behavior. In Fig. 2-12A, imagine that the pn junction has just been formed. There is, indeed, a tendency for the electrons and holes provided by the impurity atoms to migrate across the junction and terminate their existence by recombination. However, there is a secondary effect in that the impurity atoms that provide these migratory electrons and holes are left in an ionized state, bearing charges *opposite* to the departed carriers. In Fig. 2-12B we see the majority charge carriers beginning their migration toward one another. Keeping in mind that the ionized impurity atoms are locked in the crystal lattice and are therefore *not* mobile, we see an interesting pattern forming in Fig. 2-12C. Finally, in Fig. 2-12D, the impurity ions form parallel rows or "plates" of opposite charges facing each other across the depletion layer. We are reminded of a charged capacitor, with the depletion layer behaving as the dielectric. (The analogy appears to be a good one, for physics texts explain that the stored charges in a physical capacitor actually reside on the inner surfaces of the metal plates.) The strange dilemma that no voltage or current from this "charged capacitor" is available from the terminals of a practical pn diode will be explained shortly. Fig. 2-14 combines the ideas developed in Figs. 2-12 and 2-13. It shows the charge distribution in a pn diode.

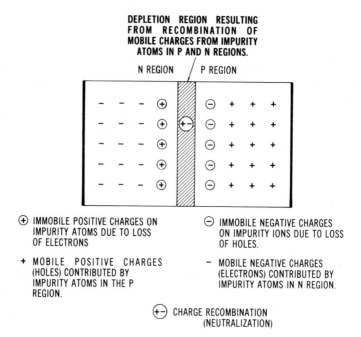

DEPLETION REGION RESULTING
FROM RECOMBINATION OF
MOBILE CHARGES FROM IMPURITY
ATOMS IN P AND N REGIONS.

N REGION   P REGION

⊕ IMMOBILE POSITIVE CHARGES ON   ⊖ IMMOBILE NEGATIVE CHARGES
   IMPURITY ATOMS DUE TO LOSS   ON IMPURITY IONS DUE TO LOSS
   OF ELECTRONS   OF HOLES.

+ MOBILE POSITIVE CHARGES   − MOBILE NEGATIVE CHARGES
   (HOLES) CONTRIBUTED BY   (ELECTRONS) CONTRIBUTED BY
   IMPURITY ATOMS IN THE P   IMPURITY ATOMS IN N REGION.
   REGION.

⊕⊖ CHARGE RECOMBINATION
    (NEUTRALIZATION)

Fig. 2-14.   The distribution of electric charges in a pn diode.

So far we do not know whether conduction takes place as the result of externally applied voltages as shown in Fig. 2-15. Let us consider the situation in Fig. 2-15B with *forward* bias applied. That is, the p region is made positive relative to the n region. The majority carriers of both regions are *repulsed* from the respective electrodes and *driven toward* the depletion layer, making it very narrow. Due to the energy acquired from the electric field, many of these carriers cross over into oppositely doped regions and are collected at the electrodes there. Although some recombination occurs in the depletion region, many of the charge carriers, which now have directed motions, pass one another, continuing their journeys until arrival in the oppositely doped region. Such arrival manifests itself as current flow in the external circuit, much as plate current in a vacuum tube results from electrons arriving at the positively charged plate.

An interesting aspect of such conduction is that the p and n regions need not be uniformly doped with their respective impurity atoms. If, for example, the diode is made with relatively heavy impurity concen-

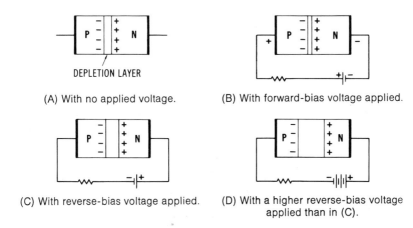

(A) With no applied voltage.

(B) With forward-bias voltage applied.

(C) With reverse-bias voltage applied.

(D) With a higher reverse-bias voltage applied than in (C).

Fig. 2-15. Behavior of the depletion region in a pn-junction diode.

tration in the p region as compared to the n region, conductivity will be primarily achieved via *holes* because of their availability. The converse situation prevails if the n region is made with relatively high impurity concentration. The conductivity, though still due to *majority* carriers, will then be primarily achieved via *electrons*. In all situations forward bias causes *majority* carriers to be driven into a narrowed depletion region, and many of these carriers *penetrate* into the oppositely charged semiconductor region to be collected at the electrode. What happens if the applied bias voltage is reversed, as depicted in Figs. 2-15C and D? If for no other reason than that we have already labeled our device a *diode,* we anticipate an entirely different phenomenon—one that *prevents* current from flowing into the external circuit.

When the p region of a pn junction is polarized negatively with respect to the n region, the mobile majority charges in the two regions are attracted to opposite charges appearing at their respective electrodes. Thus, electrons in the n region *recede* from the junction and move closer to the electrode contacting the n region. Similarly holes in the p region *recede* from the junction and move closer to the electrode contacting the p region. The mutual action of the charges in the oppositely doped regions thereby widens the depletion layer. Significantly current cannot flow by transport of majority carriers because the electric field tends to pull them *away* from the depetion layer. A very tiny current does flow due to thermally generated electron–hole pairs *within* the depletion layer. This is essentially the current that would flow in pure silicon. In summary, we have a device that provides thousands of times the conductivity in one direction of current flow than in the opposite direction. Thus, our device is a *rectifying diode*.

The Dilemma of the Internal Voltage

We have already discussed the paradox of the "charged plates" bounding the depletion layer in the pn junction. The question was raised: Why is this voltage not detectable across the ohmic terminals of an ordinary pn-junction diode? Referring to Fig. 2-16, it is entirely acceptable to assume the existence of a voltage across the junction itself. However, by the time we get out to the *electrodes*, this voltage is no longer measurable. The situation is demonstrated by the series arrangement of cells shown in Fig. 2-16B. We see that the "internal voltage" is *not* available at the outer terminals of the cascaded cells. In the pn-junction diode of Fig. 2-16A the relationship between the ions bounding the depletion layer and the majority charges in the rest of the semiconductor material is the same as that of the four cells in Fig. 2-16B.

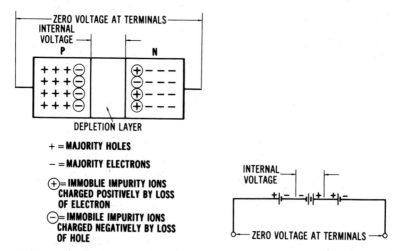

+ = MAJORITY HOLES

– = MAJORITY ELECTRONS

⊕ = IMMOBLIE IMPURITY IONS CHARGED POSITIVELY BY LOSS OF ELECTRON

⊖ = IMMOBILE IMPURITY IONS CHARGED NEGATIVELY BY LOSS OF HOLE

(A) Static representation of the charge condition within a pn-junction diode.

(B) Equivalent battery circuit showing why internal "charged-capacitor" voltage does not appear across terminals of pn-junction diode.

Fig. 2-16. Why no voltage is detectable at the terminals of a pn-junction diode.

Although this junction voltage is not available at the terminals of the pn-junction diode, it is much in evidence because the forward-bias voltage applied to the diode must be on the order of 0.6 volt before the previously described forward conduction can take place.

The Reverse Current Not Always Trivial

We have qualified the pn diode as an exceedingly good rectifying device, mainly by virtue of the high ratio of forward conduction to reverse conduction. However, certain constraints must be complied with in order for the reverse current to be truly negligible. Referring to Fig. 2-17, we see that reverse current remains substantially constant over a wide range of applied *reverse voltage*. However, reverse current is dependent on *temperature*. The dependency is strong, although this characteristic is not quantitatively revealed by Fig. 2-17. If a pn diode is allowed to operate too hot, its rectifying action will be lost or it may be destroyed. Raising the temperature of a semiconductor material increases the number of thermally generated electron–hole pairs. Because these charge pairs are free to conduct electrical current, the material changes from a good insulator at room temperature to a good conductor at elevated temperatures. Indeed, if the temperature is high enough, the conduction

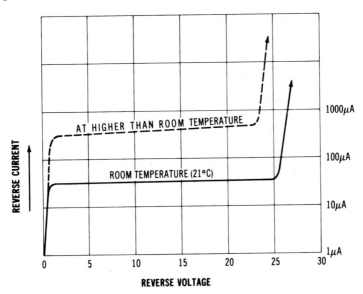

Fig. 2-17.   Reverse voltage behavior of a typical pn-junction diode.

resulting from the thermally generated charges overwhelms that resulting from the impurity atoms. Thus, the behavior of overheated silicon semiconductor material is essentially the same as for overheated intrinsic silicon.

THE PHENOMENA OF REVERSE VOLTAGE BREAKDOWN

Fig. 2-17 also indicates that a sufficiently high reverse voltage projects a pn-junction diode into an entirely different mode of conductivity, for the current abruptly undergoes a tremendous increase. For the example shown in Fig. 2-17, avalanche voltage breakdown occurs in the vicinity of 25 volts. We are reminded of the phenomenon of ionization in a gaseous device such as a neon bulb. The sudden breakdown is nondestructive as long as a series-limiting resistor is incorporated to restrict the maximum allowable current. The actual voltage at which breakdown occurs is governed somewhat by temperature but much more so by the impurity concentration in the doped regions and by various geometrical and processing factors in the structure of the diode. The breakdown voltage is thereby controlled primarily by the manufacturer. This may range from several volts to several thousand volts. Actually two different breakdown mechanisms are postulated, one known as *avalanche breakdown* and the other described as *zener breakdown*.

Avalanche Breakdown

According to this theory the thermally generated charge carriers in the depletion layer obtain sufficient energy from the accelerating force of the applied electric field to dislodge other carriers from their covalent bonds. A cumulative production of carriers ensues wherein each available carrier impacts one or more lattice-bound charges with sufficient force to overcome the covalent bond. Such liberated charges not only become current carriers but participate in the avalanche process by impacting with other lattice-bound charges. A tremendous number of carriers are created, and conduction becomes exceedingly high. This mechanism is analogous to the Townsend effect in electrified gas. The basic idea of avalanche breakdown is illustrated in Fig. 2-18. The initiation of the process is dependent on thermally generated carriers.

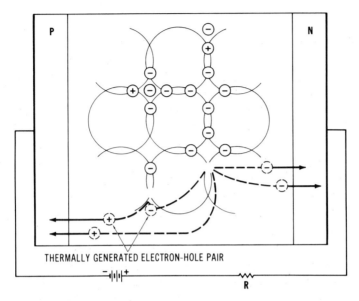

Fig. 2-18. Avalanche breakdown in a pn junction.

Zener Breakdown

A second theory, formulated by Zener, accounts for the current increase by assuming that the force of the applied electric field actually tears loose a large number of charges from their covalent bonds in the atomic structure of the crystal lattice. This process is depicted in Fig. 2-19. The difference between the two breakdown theories is somewhat subtle insofar as practical results are concerned. It has been determined, however, that the zener effect in silicon predominates at low voltages—below approximately 5.5 volts. (In germanium it appears that the avalanche breakdown occurs first. That is why there were no germanium "zener" diodes on the market when germanium was the dominant semiconductor material.)

It has become customary to designate silicon breakdown diodes as zener diodes regardless of the breakdown mechanism. More significantly every pn-junction diode, no matter what its intended function, is subject to breakdown when excessive reverse voltage is impressed

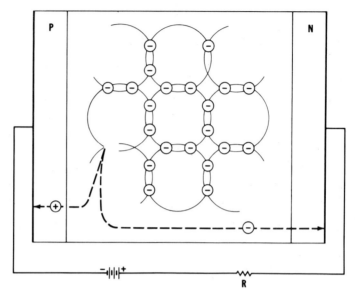

Fig. 2-19.   Zener breakdown in a pn junction.

across its terminals. And this also includes pn junctions within *other* devices, such as transistors, thyristors, and junction field-effect devices.

Reverse Breakdown with a Wide Depletion Region

It can be seen from Fig. 2-15 that the depletion layer *widens* with increasing reverse voltage. It has also been stated that the boundaries of the depletion layer are *fixed* ions. Therefore, the widening of the depletion zone cannot be due to physical movement of these parallel rows of boundary ions. What happens is that successively more distant rows of impurity atoms become boundary ions as the reverse voltage is increased. A practical proof that this actually occurs will be cited soon. But for the moment it appears strange that reverse breakdown, whether it is an avalanche or a zener mechanism, takes place in a relatively wide depletion region. If we liken the phenomenon to a spark gap in a vacuum, our confusion deepens. But the presence of gas atoms can impose voltage-breakdown characteristics in a gas tube that are not completely governed by mere physical separation of the electrodes. And such is also the case with reverse voltage breakdown in a junction diode. Despite the fact that the depletion region becomes wider as the reverse voltage is raised, a voltage ultimately reached is suddenly able to supply the energy needed to generate a "cloud" of charge carriers. As explained, this process abruptly liberates other carriers with the result that the junction becomes highly conductive.

A MATHEMATICAL VIEWPOINT
OF PN-JUNCTION CURRENT

We have accounted for both forward and reverse conduction in the pn-junction diode. One might easily suppose that entirely different equations would be needed to describe the two currents because they appear to invoke quite different conduction phenomena. In *forward conduction,* charge carriers are injected into a narrow depletion region and manage to find their way across it to oppositely doped semiconductor regions where they are "collected," thereby producing current in the external circuit. Conversely, *reverse conduction* is a kind of "stray" effect, which spoils what would otherwise be a high-quality insulating material. Specifically, thermally generated carriers within a relatively wide depletion region cause a tiny current. This current is nearly independent of the reverse voltage between about one-tenth of a volt and whatever voltage causes avalanche or zener breakdown in a particular diode. This contrasts with the *voltage-dependent* forward conduction current. Finally, reverse conduction is due to *minority* carriers only, whereas forward conduction involves both minority and majority carriers, but mainly *majority* carriers. With these differences in mind, it may come as somewhat of a surprise that a single, simple equation suffices to provide excellent quantitative data pertaining to the operation of a pn-junction diode in *both* operational modes—forward and reverse conduction.

This equation, moreover, can be deployed automatically to take into account the effect of temperature. And it applies equally well to both silicon and germanium diodes. The equation is

$$I = I_s(e^{40V} - 1)$$

where

I is the current through the pn junction
I_s is the reverse current measured at temperature T
V is the voltage across the diode

Several observations are in order. The value of temperature T need never be plugged into the equation. When V is reverse voltage, it is negative. A little experimentation reveals that for values of $-V$ numerically greater than, say, -0.1 volt, the equation becomes for practical purposes $I = -I_s$ (recall that $e^{-x} = 1/e^x$). Finally, it can easily be ascertained that *positive* values of V above a few tenths of a volt will cause I to increase at an exceedingly high rate, at which time we have *forward* conduction.

This single equation is quite oblivious of the voltage where avalanche or zener breakdown occurs in any given pn junction. It is able to tell us only that current I will ideally be approximately I_s over a range of reverse voltage. However, the reverse breakdown voltage is ordinarily available from the manufacturer's specifications or from the direct measurement.

THE JUNCTION TRANSISTOR

As its name implies, the junction transistor derives its function from the behavior of the pn junction. Also implied by its name, the operation of the junction transistor involves the transit of a current from one region to another. Such an implication may appear to be trivial. However, it is well to bear in mind that if by some means a given current can be made dependent on a *low* resistance, and then passed through a *high* resistance, *power amplification* can be achieved in the high resistance. From Ohm's law, a small change in current in the low resistance, R_1, will be accompanied by a power variation, P_1. This change in current will evoke a *larger* power variation, P_2, in the high resistance, R_2. This follows from $P = I^2R$. Unfortunately, if we merely connect two resistances in series in an attempt to prove this fact, our endeavor is defeated because the circuit current is primarily governed by the *high,* rather than the low, resistance. For the moment we are in possession of a tantalizing hypothesis, but one seemingly predicated more on fancy than fact. Obviously there would be no point in indulging in fantasy—so let us investigate how actual realization of this power-gain mechanism is brought about in the transistor.

The transistor structure illustrated in Fig. 2-20A is seen to consist of two junction diodes fabricated in a single pellet of semiconductor material. It is shown connected to dc bias sources in such a manner that

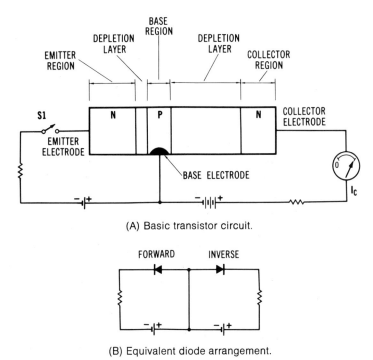

(A) Basic transistor circuit.

(B) Equivalent diode arrangement.

Fig. 2-20. Basic transistor structure and equivalent back-to-back diode arrangement.

the left-hand diode is forward biased and the right-hand diode is reverse biased. This is inferred not only by observation of battery polarity but also by the relative widths of the depletion layers. The base region electrode is *common* to both "diodes." This arrangement is suggestive of the discrete diode circuit shown in Fig. 2-20B. A major *difference* between these two circuits is that in Fig. 2-20A, as we shall see, the current in the right-hand "diode" is much determined by the current in the left-hand diode. In Fig. 2-20B there is *no* interdependency between the two diodes. Although the circuit in Fig. 2-20B depicts *some* aspects of the circuit in Fig. 2-20A, the resemblance proves superficial if we wish to simulate the power amplification possibilities of the transistor by the discrete diode circuit.

If, as has been inferred, the left-hand diode is capable of influencing the right-hand diode, it follows that the left-hand diode must behave as some sort of "charge gun," which injects current carriers into the right-hand diode. We might suspect such a mechanism by virtue of the reverse-bias voltage applied to the collector-base (right-hand) diode. If its behavior stemmed merely from its own characteristics, the collector-base diode would support only a tiny current in its external circuit and we would in essence have the back-to-back diode circuit of Fig. 2-20B.

THE SEMICONDUCTOR ELECTRON GUN

If we consider the emitter-base diode of Fig. 2-20A by itself, its function would be suggestive of a thermionic electron gun such as might be found in a cathode-ray tube. This is illustrated in Fig. 2-21. The qualitative analogy is good enough to provide further insights. However, the electrons emerging from the accelerating grid or electrode in the thermionic gun have accelerations imparted by an electric field. This is not so in the semiconductor version, because the electrons that make their way across the base region are subjected to a virtually zero electric field. What impels their transit more or less horizontally across the base region, and why aren't they all "collected" at the bottom base electrode? The answers to these two questions are of key importance to the implementation of a transistor structure.

Note the zig-zag path indicated for the electrons in the p-type base region in Fig. 2-21A. This implies that electronic movement is due to *thermal agitation* and to the development of a so-called *concentration gradient*. A concentration gradient arises from the fact that like charges repel, causing a tendency for the electrons continually to seek regions of the base material in which electron population is less than in their present region. Such a mechanism helps *guide* the net movement of electrons across the thin base region. In other words the electrons have a directed transverse movement *superimposed* on their random motions. And because the p-type base region is purposely made thin, most electrons arrive at the far end region of the p-type material rather than at the base electrode. Such electrons are then susceptible to the attractive force of the collector electrode. (If it is not strictly proper to say that these electrons are "shot" into the collector region of the transistor structure, we can at least say that they are "introduced" there.)

You might suspect that this *largely haphazard migration* of electrons must limit the speed of response of the transistor. This indeed is true. However, processing technology has advanced the frequency range

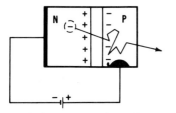
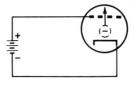

(A) Simplified representation of an emitter-base diode as a charge carrier.

(B) Analogous phenomena in a positive-grid vacuum tube or electron gun.

Fig. 2-21. The base-emitter diode can be considered analogous to an electron gun.

from tens of kilohertz in early power transistors to hundreds of megahertz in some modern designs. Some special microwave power transistors extend frequency response into the thousands of megahertz. As expected, one of the ways in which greater frequency response has been attained has been through techniques for forming *thin* base regions.

By locating the base electrode on the side of the semiconductor "gun" as shown in Fig. 2-21A, the electrons are made to *enter* the base region but are for the most part *not* collected by the base electrode. Another stratagem is involved in making the impurity doping in the n-type emitter region *much heavier* than in the p-type base region. This causes the predominant portion of charge movement to consist of electron injection from the emitter to the base region. Only a relatively small portion of charge movement consists of hole injection from the base to the emitter region. Electron collection by the base electrode tends to defeat the objective of the "gun" to make electrons available for the electric field in the collector region. Hole injection from the base to the emitter is also counterproductive for it serves only to increase external emitter base biasing current.

It is not to be inferred that such an electron gun as just described would work if placed in a vacuum. It certainly does function, however, when made an *integral part of the transistor structure.* We have considered it in "detached" form only for the sake of simplifying the discussion.

An interesting aspect of charge injection into the collector region of the transistor is that the charges coming out of the emitter region are *majority* carriers in *that* region. That is, they are the carriers resulting from impurity atoms in the emitter region. For the npn transistor structure under discussion, these charges are electrons and they become *minority* carriers in the base region because of the p-type semiconductor material. When these electrons enter the n-type collector region, they again constitute *minority* carriers. That is, they are the same type of charges that *already* are producing a small reverse current in the collector-base diode. It will be recalled that minority charge carriers are generated by thermal agitation and give rise to a relatively small current at room temperature. (Current due to minority carriers in any semiconductor region is simply the current that would exist without impurity atoms. Even though diode regions are doped with impurity atoms, reverse biasing produces only that current arising from thermal generation of *minority* carriers.)

From the foregoing discussion we have the interesting situation of a reverse-biased base-collector diode in which the tiny "natural" reverse current can be *greatly increased* by introducing a copious supply of current carriers from the emitter-base "electron gun." The prospects for amplification and power control now assume feasibility.

COMPARISON OF TRANSISTOR
AND VACUUM-TUBE AMPLIFIER

Our discussion of processes involved in junction transistor action is probably the minimum needed by those whose primary interest is solid-state circuits and applications. It should be obvious that those wishing to specialize in the physics and chemistry of semiconductors should study the comprehensive texts devoted to this phase of the technology. We shall now culminate our investigation of transistor action by directing attention to similarities existing in the transistor and vacuum-tube amplifiers.

A comparison between transistor and vacuum-tube amplifiers is shown in Fig. 2-22. The npn transistor is depicted in a common-emitter circuit because this is the most popular circuit configuration. Another reason for dealing with the common-emitter circuit is because this is the way vacuum-tube amplifiers are most often used. The control grid of the tube is positively biased. This is done in order to reinforce the analogy and should prove acceptable inasmuch as it is permissible to operate tubes in this fashion under certain circumstances. For example, so-called "space-charge" tubes once used in automobile radios operated approximately this way. Also the control grids of Class B amplifier tubes are often driven into their positive regions. Actually, when a vacuum

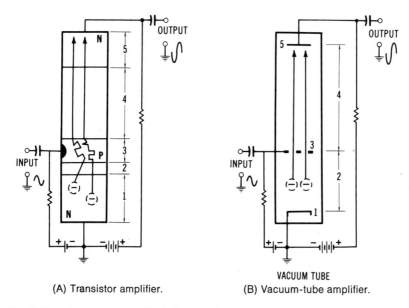

(A) Transistor amplifier.   (B) Vacuum-tube amplifier.

Fig. 2-22. Comparison of the behavior of a transistor amplifier and a vacuum-tube amplifier.

tube with the control grid positive with respect to the cathode is performing as an amplifier, the resemblance to transistor action is surprisingly close because there is grid-cathode current corresponding to base-emitter current in the transistor. Also the input impedance of such a tube amplifier is relatively low, as is that of the transistor.

Table 2-1 will help correlate similar phenomena in the two devices shown in Fig. 2-22. Further insights may be gleaned from the analogies shown in Fig. 2-23. Here, not only are corresponding currents the same, but in both devices the current in the "input" diode is made up of negative charges flowing in one direction and positive charges flowing

TABLE 2-1. Corresponding Regions within Transistors and Tubes

Transistor Region	Function or Property	Vacuum-tube Region
Emitter region	Charges originate	Emitter or cathode
Depletion layer	Electric field region	Emitter-grid region
Base region	Control region	Grid
Depletion layer	Electric field region	Plate-grid region
Collector region	Charges collected here	Collector or plate

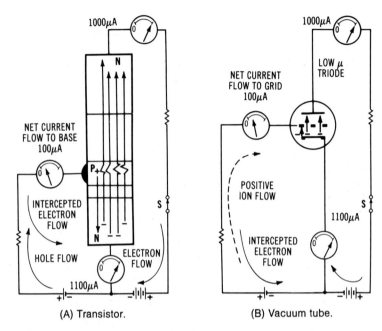

(A) Transistor.   (B) Vacuum tube.

Fig. 2-23. A more detailed example of analogous operation of a transistor and a vacuum tube.

in the opposite direction. Significantly in both cases the *net* current measured in the control-electrode circuit is relatively small, being comprised of the *difference* between the currents in the collector (plate) and emitter (cathode) leads.

THE SILICON CONTROLLED RECTIFIER

The development of the bipolar transistor from the basic pn-junction diode appears natural enough in retrospect. But surely imagination and perhaps more than a trivial measure of luck were involved. The same has been true for *another* derivative of the pn junction—the silicon controlled rectifier (SCR). This time one had the advantage that bipolar transistors could be used as building blocks for a useful new device. Consider the simple transistor circuit of Fig. 2-24A. The outputs of the transistors, one an npn and the other a pnp, are connected to the inputs of one another. This arrangement is no doubt familiar enough because

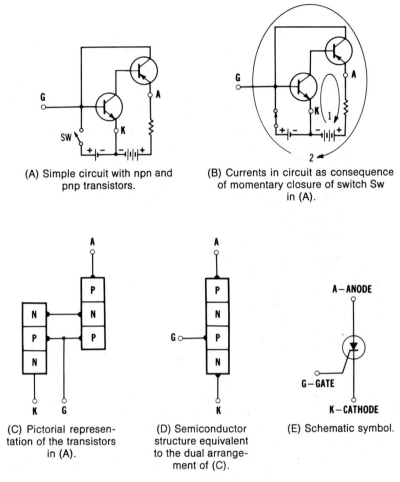

(A) Simple circuit with npn and pnp transistors.

(B) Currents in circuit as consequence of momentary closure of switch Sw in (A).

(C) Pictorial representation of the transistors in (A).

(D) Semiconductor structure equivalent to the dual arrangement of (C).

(E) Schematic symbol.

Fig. 2-24. Building the silicon controlled rectifier.

the first applications for the early junction transistors were in various oscillator and multivibrator circuits.

In any event a momentary closure of switch Sw leads to a regeneratively reinforced latch-up condition with two current paths as depicted in Fig. 2-24B. The two transistors are turned on and remain heavily saturated because one current path begets the other. Indeed this cause-and-effect relationship is so persuasive that no signal applied to G, the gate of the circuit, can turn the transistors off. Only by interrupting the power-supply current can the transistors be restored to their "off" conductive state. It is apparent that we have devised a good latching switch.

Fig. 2-24C is a pictorial representation of the transistors involved in the circuit of Fig. 2-24A. A *single* equivalent structure that can be made to substitute for this dual arrangement is shown in Fig. 2-24D. This consists of *four* semiconductor regions and *three* pn junctions. This is the basic structure of many thyristor devices including the SCR. Thus when fabricated on a single "chip," the thermal and electrical properties assume favorable values, and the economic and practical aspects also become attractive.

As a regenerative switch, the SCR and similar thyristors have revolutionized solid-state power electronics. When power is controlled by various switching techniques employing these devices, efficiency is high because such switches are either on or off—never in intermediate or "linear" states where high power dissipation occurs. Of course, the circuit designer must exercise some ingenuity at times in order to control the duration of the on time and to commutate, or turn off, such devices in a desired fashion.

Fig. 2-25 illustrates the electrical characteristics of the SCR as a two-terminal device, that is, with no current delivered to the gate. Although the SCR is not commonly employed in this fashion, it provides insight to consider it first in this simple operating mode. Also one would have to use the basic information in such an operating curve in order to specify an appropriate SCR for given circuit conditions. For example,

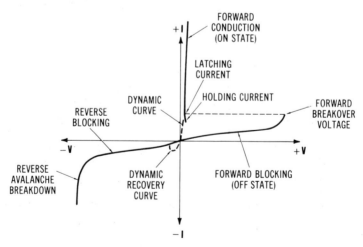

Fig. 2-25. Electrical characteristics of an SCR with zero gate voltage.

if the circuit voltage equals or exceeds the forward breakdown voltage, the regenerative switching will take place without any gate signal. Also the reverse voltage cannot equal or exceed the avalanche breakdown value. Otherwise the SCR will no longer be an open switch for reverse voltage and will be forced to dissipate high, perhaps destructive, power. In addition the load must permit sufficient holding current so that the SCR will remain conductive once triggered.

The effect of gate current is depicted in Fig. 2-26. The basic information presented in these curves is that *sufficient* gate current (at least I_{G4}) should be delivered to cause the SCR to switch abruptly from the blocking, or off, state to the on state over the entire anode-cathode voltage range. If the gate current is too small, the SCR will *not* be triggered at low voltages. Sufficient gate current is also desirable to promote fast triggering, to provide a safety factor for the effects of temperature, and to ensure essentially similar circuit operation with different SCRs. The latter objective is generally met by using a trigger device such as a neon lamp, or preferably a semiconductor device such as a diac, a silicon unilateral switch, or a UJT.

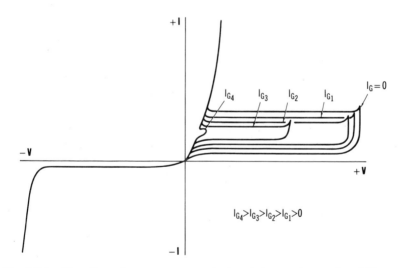

Fig. 2-26.   The effect of gate current on the breakdown voltage of an SCR.

An example of an SCR in a phase-control circuit is shown in Fig. 2-27. Here the speed of a universal motor is controlled by varying the rms current through its armature and field winding. Although such

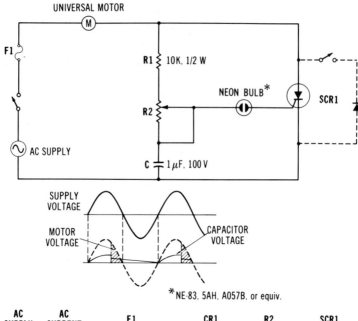

Fig. 2-27. Example of a simple SCR motor-control circuit.

control is efficient and effective for many applications, it should be borne in mind that top speed is essentially that which would prevail with the motor connected to the ac line through an ordinary silicon rectifying diode. If such a limitation is not allowable, the diode-switch circuit shown in dashed lines can be added. Ordinarily the switch should be closed only when the SCR is operating at full conduction (approximately 180 degrees). However, the best way to achieve wide range continuous control is to use back-to-back SCRs or a single triac.

Half-wave power control of motors, heaters, and lamps continues to merit consideration, in spite of the development of the full-wave thyristor, the triac. This is largely because SCRs remain superior when it comes to high power capability.

THE TRIAC

Although an SCR provides satisfactory control of ac power for many applications, there are few instances where full-wave rather than half-wave, operation would not be desirable. In the full-wave mode it is easier to obtain a wide range of power control. Even when the load is resistive, such as a heater or lamp, the power factor presented to the ac line is better (higher) with full-wave control. Moreover, if an isolating transformer is used, half-wave operation tends to produce core-saturation problems due to the dc component associated with this mode. Such an effect is not encountered with full-wave control. In certain inverter and power supply applications, filter problems are reduced in full-wave operation because the lowest harmonic present in the output is 2f, rather than f as with half-wave operation (f is the frequency of the ac voltage applied to the thyristors). Also, all *odd* harmonics except f, the *first harmonic,* are cancelled in the output of a full-wave control circuit.

A feasible method of utilizing SCRs in full-wave control systems is shown in Fig. 2-28A. Such a back-to-back connection is quite straightforward. One SCR is phase-controlled for one-half of the ac cycle and the other SCR operates in symmetrical fashion for the alternate half-cycle. This results in full-wave control of load current. There is one drawback to this scheme insofar as circuit simplicity in that the gates cannot be tied together. Rather some isolation technique must be used so that the gate of each SCR can be triggered with respect to *its own* cathode. Additionally the phase of the trigger signals must be displaced by 180 degrees even though their individual phases are varied in order to provide control of the load power. Finally, the two trigger-signals should be alike in amplitude, source impedance, rise time, etc. All of these requirements are readily met via the method shown in Fig. 2- 28B. The gate control circuit is often a UJT relaxation oscillator.

Despite the feasibility of the basic scheme represented by Fig. 2-28B, it would be even better if a bidirectional thyristor could be devised, particularly one with a *single* control gate. Intuitively one might suspect that such a device would be possible because a given semiconductor geometry could be "shared" in time—in full-wave operation only one SCR in the circuits of Fig. 2-28 carries load current at any one time. And inasmuch as the two SCRs are identical, why not utilize a *single* structure for such a control purpose? Indeed such reasoning led to the

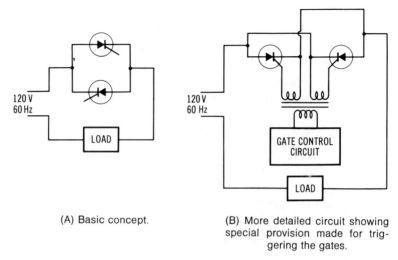

(A) Basic concept.

(B) More detailed circuit showing special provision made for triggering the gates.

Fig. 2-28.   The use of back-to-back SCRs to achieve full-wave control of load power.

development of the *triac,* an acronym for *tri*ode *ac* switch. It is also known as a *bidirectional triode thyristor.*

The symbol for the triac shown in Fig. 2-29A is intended to depict back-to-back SCR elements with a single-gate electrode. In order to bring about the single-gate feature, the structure of the triac entails greater complexity than would result from simply combining two SCR structures. The terminals that handle the principal current through the device are labeled *main terminal 1* (MT1) and *main terminal 2* (MT2) rather than anode and cathode as in the SCR. The reason for this can be gleaned from inspection of the voltage-current characteristics of the device that are shown in Fig. 2-29B. Because there is no reverse voltage, symmetrical performance prevails for both polarities of applied voltage. This feature, of course, is needed for full-wave operation. Accordingly, terminal MT1 is arbitrarily considered the "reference" terminal; the polarities of the gate and terminal MT2 are always designated with respect to MT1.

It is also true that the triac can be triggered by *either* a positive or a negative gate signal, whether terminal MT2 is positive or negative. Triacs are available with different gate sensitivities in order to accommodate various triggering modes. Many circuit applications favor trig-

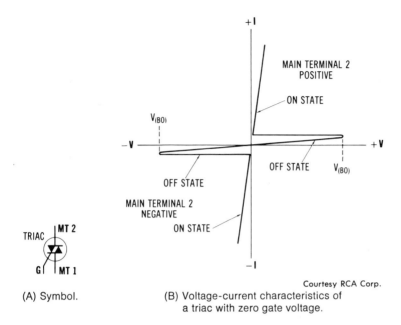

(A) Symbol.

(B) Voltage-current characteristics of
a triac with zero gate voltage.

Courtesy RCA Corp.

Fig. 2-29. The triac is a full-wave thyristor which substitutes for a pair of SCRs.

gering with a positive gate signal when terminal MT2 is positive and
with a negative signal when terminal MT2 is negative.

The same need for gate trigger signals of *sufficient amplitude* as
described for the SCR also applies to the triac. Again triggering will
not occur at lower device voltages if sufficient gate current is not available. With sufficient gate current the triggered triac will regeneratively
revert from the blocking state to the conductive state. Once turned on
in this manner, its characteristics resemble those of a forward-biased
silicon diode.

The triac is predominantly used in ac circuits, and commutation
occurs "naturally" as the current through terminals MT1-MT2 reverses
direction. Triac operation is shown in Fig. 2-30 for the four possible
combinations of polarity for terminal MT2 and the gate polarities. Figs.
2-30A and D correspond to the mode of operation usually encountered
in phase-control circuits. Note that the semiconductor "sandwich" is
pnpn downward from terminal MT2 for one direction of the principal
current and is npnp downward from terminal MT2 when the principal
current is in the opposite direction.

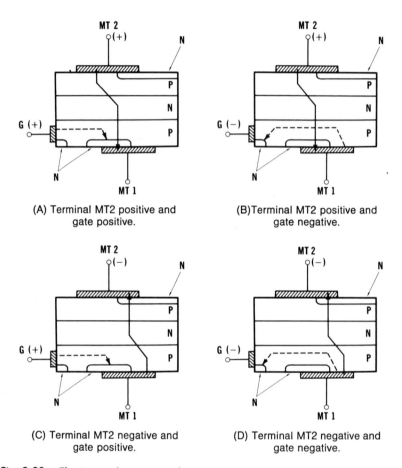

(A) Terminal MT2 positive and gate positive.

(B) Terminal MT2 positive and gate negative.

(C) Terminal MT2 negative and gate positive.

(D) Terminal MT2 negative and gate negative.

Fig. 2-30. The internal current paths in a triac.

A typical circuit application of a triac is shown in Fig. 2-31. Besides providing wide-range full-wave control, this circuit features a dual-time-constant phase-shift network to discourage hysteresis in the setting of R2, a diac triggering device, and an rfi filter.

THE JUNCTION FIELD-EFFECT TRANSISTOR

Some of the hardware applications of the pn junction described have included the rectifying diode, the transistor, the varactor diode, the zener diode, and the thyristor. There is at least one other important

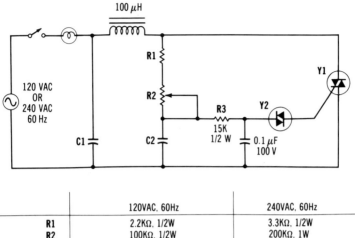

	120VAC, 60Hz	240VAC, 60Hz
R1	2.2KΩ, 1/2W	3.3KΩ, 1/2W
R2	100KΩ, 1/2W	200KΩ, 1W
C1 C2	0.1μF, 200V	0.1μF, 400V
Y1	RCA 40485 ⎱ or 40431	RCA 40486 ⎱ or 40432
Y2	RCA 40583 ⎰	RCA 40583 ⎰

Courtesy RCA Corp.

Fig. 2-31.   Typical triac lamp-control circuit.

implementation of the pn junction. Until recently the junction field-effect transistor (JFET) was found only in small-signal and low-power applications. However, these devices, along with insulated-gate FETs (MOSFETs), are serving as output devices in stereo amplifiers, in electric motor controls, and in other power systems requiring control of tens or hundreds of watts. Insight into the principle of the junction FET is readily attained by recalling the width modulation of the depletion layers in varactor diodes and in transistors. If one can control the geometry of the depletion regions by a voltage (as represented by an electric field), it should be possible thereby to vary resistance to a current. The situation should be somewhat analogous to changing the cross-sectional area of a conducting wire. Unfortunately the structures of diodes and transistors do not readily lend themselves to such behavior.

Fig. 2-32 illustrates the basic concept of a simple junction FET. The idea is to provide a length of semiconductor material with a central region where construction of the effective cross-sectional area can take place. The means of producing such a constriction is by using a reverse-biased pn junction. The control voltage of the device is applied to one side of the pn junction through a terminal appropriately known as the

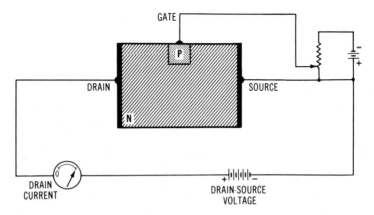

(A) Low reverse bias applied to gate.

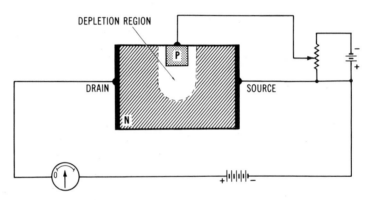

(B) Moderate reverse bias applied to gate.

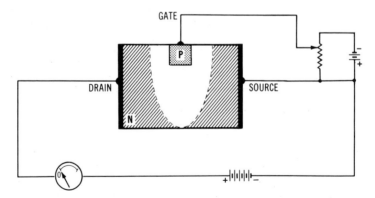

(C) High reverse bias applied to gate.

Fig. 2-32. Basic operating principle of the junction FET.

gate. Note that the load current is controlled by the effective ohmic resistance of a path through the semiconductor material and does not pass through any pn junction, as in a bipolar transistor. (Accordingly this device has been called a *unipolar transistor.*) Recent versions of FETs intended for power applications are fabricated to cause the controlled current to follow vertical, rather than horizontal, paths. However, the basic operating principle remains essentially the same.

THE INSULATED-GATE FIELD-EFFECT TRANSISTOR

Inasmuch as the gate junction of the JFET power transistor is actually deployed as a capacitor, it would be only natural to consider the feasibility of using an actual physical capacitor as a means of introducing an electric field into the semiconductor channel. Such a device is indeed practical as shown by the general configuration of the insulated-gate FET (IGFET or MOSFET) illustrated in Fig. 2-33. The gate capacitor is comprised of a film of silicon dioxide as the dielectric, and a metallized layer as one "plate"—the other plate may be said to be the p-type semiconductor material of the substrate. The electrical action of this device can be seen to be similar to that of the previously described JFET. Along with junction field-effect transistors, the emergence of insulated-gate types as power-controlling devices is relatively new. This may be somewhat surprising to those who have been conditioned for many years to regard all FETs as inherently limited to signal-level applications.

Both the JFET and MOSFET thus far discussed operate in the *depletion mode.* That is, they conduct heavily without bias voltage applied between the gate and source. The application of bias voltage then reduces the current through the drain-source circuit. The MOSFET (but not the JFET) can also be made so that operation occurs in the *enhancement* mode. In such operation, forward-bias voltage must be applied between the gate and source in order to produce current in the drain-source circuit. This, of course, is similar to the situation prevailing in ordinary bipolar transistors. A significant *difference* is that the input circuit of the bipolar transistor consumes current, whereas the gate of the MOSFET makes no such demand from the input source. Indeed input resistance of the MOSFET can be about 10^{15} ohms, which is even higher than that of the JFET gate.

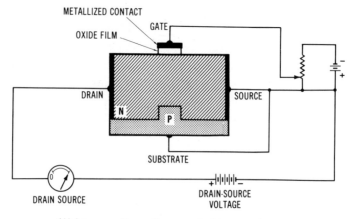

(A) Low negative voltage applied to gate electrode.

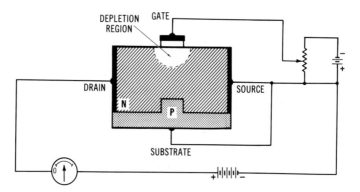

(B) Moderate negative voltage applied to gate electrode.

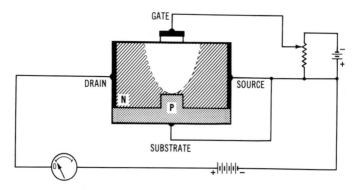

(C) High negative voltage applied to gate electrode.

Fig. 2-33. Basic operating principle of the depletion-mode MOSFET.

The principle of the enhancement-mode MOSFET is illustrated in Fig. 2-34. The operating mechanism is somewhat more complex than in the depletion-mode device. The lower "plate" of the internal capacitor is again the p-type material of the substrate. The fabrication of the device is such, however, that this p-type region interrupts the n-type current-carrying channel. Accordingly, without gate voltage, drain-source current is negligible. The application of positive voltage between the gate and source *induces* an n-type region in the p-type material, thereby producing a controllable current-carrying n-type channel as shown in Figs. 2-34B and C.

It may appear that there is variance in the effect of the electric field in the depletion-mode and in the enhancement-mode devices. In one case the semiconductor material becomes "depleted"; in the other case an *induced* n-type region appears. The two effects are, however, manifestations of similar phenomena and can be manipulated by the manufacturer when impurities are "blended" with the intrinsic silicon.

P-CHANNEL FETs

Hitherto we have considered only n-channel FETs. Field-effect transistors are, as with bipolar transistors, available with two types of polarity. The n-channel devices correspond to the npn transistor insofar as concerns bias polarities, and p-channel FETs correspond to pnp transistors. Fig. 2-35 shows p-channel versions of the junction and insulated-gate FETs previously discussed.

In the simple examples presented, it might appear that symmetry prevails with respect to source and drain connections and that they can be interchanged. This was true with some small-signal devices but is not generally the case with the power FETs, which have asymmetrical structures because the substrate is internally connected to the "source" and because of integrally contained protective diodes.

The V-groove, or "vertical" type, MOSFET used in power devices is shown in Fig. 2-36. This structure operates in the enhancement mode. The drain region encompasses a large area and permits effective heat removal. A protective zener diode for the gate is monolithically fabricated in this structure.

The different types of field-effect transistors are summarized and compared to bipolar transistors in the block diagrams of Fig. 2-37.

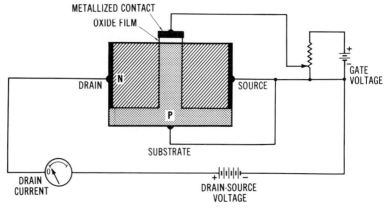

(A) Low positive voltage applied to gate electrode.

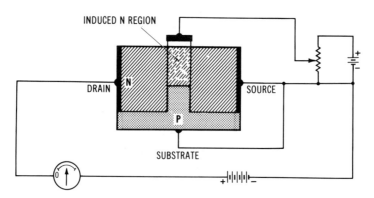

(B) Moderate positive voltage applied to gate electrode.

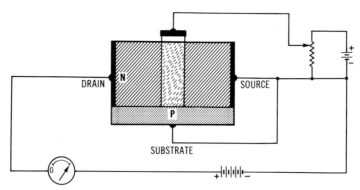

(C) High positive voltage applied to gate electrode.

Fig. 2-34. Basic operating principle of the enhancement-mode MOSFET.

96

Fig. 2-35. P-channel versions of junction and insulated-gate FETs.

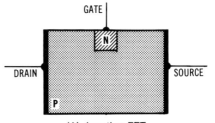

(A) Junction FET.

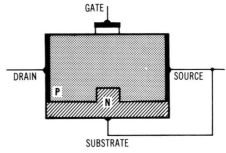

(B) Depletion-mode MOSFET.

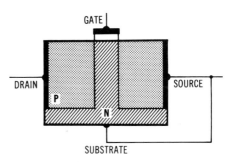

(C) Enhancement-mode MOSFET.

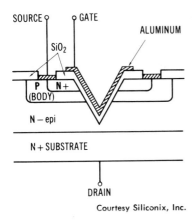

Courtesy Siliconix, Inc.

Fig. 2-36. Geometry used in V-grove power MOSFETs.

97

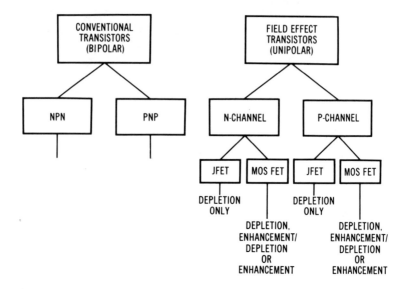

Courtesy Siliconix, Inc.

Fig. 2-37. Comparison between bipolar transistors and field-effect transistors.

CONSIDERATION OF THERMODYNAMICS

When it comes to the design and implementation of solid-state power systems, the "thermal circuit" often requires the greatest consideration. The matter of heat removal is so intimately related to performance, reliability, size, and cost that any attempt at design or evaluation solely on the basis of electrical considerations is destined to be misleading, to say the least. At the same time it would be quite difficult to provide specific guidance for the thermal management of each of the many practical applications to be presented later in this book. This is because of the great variation in individual construction and installation practices. Probably the most practical approach is to develop an awareness of the principles of heat flow as they apply to individual situations.

The reference to the *thermal circuit* is more than a mere play on words. Indeed we make use of a so-called Ohm's law in the selection of power devices, in the methods of mounting them, and in the techniques of heat removal. The basic relationship, which appears simple enough, is given by the equation

$$\theta = \frac{T}{P}$$

where

> θ represents the quantity known as *thermal resistance* which is expressed in degrees centigrade per watt (°C/W)
>
> T is the temperature rise in centigrade with respect to some reference temperature such as the ambient temperature
>
> P is either the allowable or resultant power dissipation in watts

Of course, the algebraic permutations of this equation are equally useful. They are $P = T/\theta$ and $T = P\theta$.

Because of our natural liking for analogies, we welcome such an easy entry into the domain of thermodynamics. A closer examination of the simple relationship $\theta = T/P$ may invoke second thoughts, because we reflect that the electrical counterpart is $R = E^2/P$. From this it must be inferred that temperature T is analogous to voltage squared. Not only do we feel ill at ease with such a strange entity, but it is apparent that the Ohm's law of the thermal circuit is not really a one-to-one equivalent of the familiar electrical relationship. The explanation of this strange situation is not likely to be found in electronic literature and requires considerable effort to dig it out of most physics textbooks.

It turns out that much use is made of thermal *conductance* in thermodynamics. As would be suspected, thermal conductance is $1/\theta$, or the reciprocal of thermal resistance. Thermal conductance is appropriately defined as the *flow rate of thermal energy per degree of temperature*. If we express this statement mathematically, we should arrive at a clearer understanding of the way in which the so-called Ohm's law of the thermal circuit comes into being.

Energy in an electrical, thermal, or any other system can be expressed as the product of power and time, or $P \times t$. Using the definition given, we can write our equation for thermal conductance:

$$\frac{1}{\theta} = \frac{\dfrac{P \times t}{T}}{t}$$

If we did not already have the definition for thermal conductance, an inspection of this equation would indeed tell us that thermal conductance is the *flow rate of thermal energy per degree of temperature*. Having thus derived this equation to comply with a basic concept in physics, we can simplify it by cancelling the *t*'s, leaving

$$\frac{1}{\theta} = \frac{P}{T}$$

Finally, if we invert this expression for thermal conductance, we arrive at our original Ohm's law relationship for thermal *resistance;* that is, $\theta = T/P$.

In the thermal circuit shown in Fig. 2-38 the resistances represent the thermal resistances associated with a power transistor and its heat sink. Note that power in the form of heat input is depicted as circulating in these resistances and that certain temperatures exist at discrete points in the circuit. This makes the determination of important thermal quantities in a practical situation quite straightforward. Needed information is generally available from four sources: manufacturer's specifications sheets for solid-state devices, various tables and charts, electrical computations, and thermal Ohm's law computations. The usual procedure is best illustrated by a typical example.

In Fig. 2-38, θ_{JC} represents the thermal resistance from junction to case. This is found in the specification sheet and there is nothing we can do to change it. Here θ_{CS} represents the thermal resistance from case to heat sink, and θ_{SA} represents the thermal resistance from heat sink to ambient. The general objective is to make *these* two thermal resistances as low as skill, cost, and packaging considerations will permit. (When a heat sink is not used, θ_{CA}, the thermal resistance from case to ambient, assumes great importance. This thermal resistance is also often extracted from the manufacturer's specifications for the device.)

Suppose that we are dealing with a voltage-regulated power supply using an RCA RCS29 power transistor as the series-pass element. Assume circuit conditions are such that maximum power dissipation occurs when this transistor is passing 2.25 amperes and has a voltage drop of 5 volts. The 11.25 watts of electrical power thereby dissipated is also 11.25 watts of "heat power." A destructive rise in junction temperature will occur unless a heat-removal technique is used. The mechanical and dimensional design of the power supply suggests that a heat sink in

Fig. 2-38. Thermal circuit for a power transistor mounted on a heat sink.

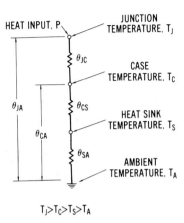

$$T_J > T_C > T_S > T_A$$

freely convected air (no fan or blower) might be suitable. How do we specify such a heat sink? The following steps typify the general procedure:

1. T_J and θ_{JC} are recorded from the device specifications. For the RCS29 a junction temperature, T_J, up to 200°C is indicated, and the "built-in" junction-to-case thermal resistance, θ_{JC}, is 5.8°C/W.

2. From Fig. 2-38 we can set up our thermal equation as follows:

$$\theta_{JC} = \frac{T_J - T_A}{P}$$

In this way we shall solve for the total thermal resistance, θ_{JA}, in the circuit under operating conditions. (This is a necessary step in arriving at the thermal resistance of *one portion* of the circuit, specifically θ_{SA}, which is the heat-sink-to-ambient thermal resistance.)

In making the numerical substitutions it is usually best to use less than the maximum allowable junction temperature. In our example we shall limit T_J to 175°C. Then, assuming a temperature of 25°C for T_A, we have

$$\theta_{JA} = \frac{175 - 25}{11.25} = 13.3°C/W$$

3. If we next obtain the two thermal resistances, θ_{JC} and θ_{CS}, we will be able to compute the maximum allowable thermal resistance from heat sink to ambient (θ_{SA}). It is then a simple matter of perusing through heat-sink catalogs to select a heat sink which is specified for θ_{SA}, or a lower thermal resistance. Table 2-2 shows commonly encountered values of θ_{CS} for power transistors with the popular TO-66 and TO-3 metal cases. Our RCS29 transistor utilizes the TO-66 case. Assuming that a 2-mil mica insulating washer is to be used with a heat-sinking compound, it is seen that θ_{CS} is in the vicinity of 2.3°C/W.

4. From $\theta_{SA} = \theta_{JA} - \theta_{JC} - \theta_{CS}$, we find that $\theta_{SA} = 13.3 - 5.8 - 2.3 = 5.2$°C/W. Referring to Table 2-3, we can choose from a number of heat sinks made by Wakefield, Thermalloy, IERC, and Staver. All of these manufacturers feature models with heat-sink-to-ambient thermal resistances in the 3.0 to 5.0°C/W range. For example, one could select the Thermalloy type 6606.

When fans or blowers are employed to provide forced air rather than natural convection, it is roughly equivalent to lowering the heat-sink-to-ambient thermal resistance, θ_{SA}. Although the improvement thereby realized is often appreciable, it is not always easy to calculate the effects of providing moving air. Much depends on the nature of the motion (whether laminar or turbulent) and the air velocity in the immediate vicinity of the cooling fins on the heat sink. One can make use of the graphs and empirical data that abound in technical literature, but usually good fortune must prevail in order to adapt such information to a particular thermal situation. It is generally best to consult handbooks and manufacturers' literature for guidance, but be prepared for a good measure of experimental investigation.

In any situation, cooling occurs not only via conduction and convection but through *radiation* as well. It is usually rewarding to enhance this means of heat removal. This is particularly true because a prime consideration of radiation involves the nature of exposed surfaces. Otherwise knowledgeable people sometimes associate radiating efficiency with a bright and shiny surface. They probably think of electric heaters with mirrorlike reflectors. Such erroneous judgment stems from the overlooked fact that the heat is not developed *within* the metallic mass of the heater reflector as it is in a transistor and its heat sink. A so-called black body constitutes the ideal radiator, but a rough, dark surface often

TABLE 2-2. Typical Case-to-Heat-Sink Thermal Resistances for Transistors with TO-3 and TO-66 Packages

Case	θCS			
	Metal-to-Metal		Using an Insulator	
Case	Dry	With Heat-sink Compound	With Heat-sink Compound	Type
TO-3	0.2°C/W	0.1°C/W	0.4°C/W 0.35°C/W	3 mil MICA Anodized Aluminum
TO-66	1.5°C/W	0.5°C/W	2.3°C/W	2 mil MICA

Courtesy Motorola Semiconductor Products, Inc.

TABLE 2-3. Heat-Sink-to-Ambient Thermal Resistance, θ_{SA}, for Some Commercially Available Heat Sinks

	TO-3 & TO-66
$\theta SA*(°C/W)$	Manufacturer/Series or Part Number
0.3-1.0	Thermalloy — 6441, 6443, 6450, 6470, 6560, 6590, 6660, 6690
1.0-3.0	Wakefield — 641 Thermalloy — 6123, 6135, 6169, 6306, 6401, 6403, 6421, 6423, 6427, 6442, 6463, 6500
3.0-5.0	Wakefield — 621, 623 Thermalloy — 6606, 6129, 6141, 6303 IERC — HP Staver — V3-3-2
5.0-7.0	Wakefield — 690 Thermalloy — 6002, 6003, 6004, 6005, 6052, 6053, 6054, 6176, 6301 IERC — LB Staver — V3-5-2
7.0-10.0	Wakefield — 672 Thermalloy — 6001, 6016, 6051, 6105, 6601 IERC — LA, uP Staver — V1-3, V1-5, V3-3, V3-5, V3-7
10.0-25.0	Thermalloy — 6013, 6014, 6015, 6103, 6104, 6105, 6117

Courtesy Motorola Semiconductor Products, Inc.

performs well. The radiating efficiency, or emissivity, of some practical surfaces is given in Table 2-4. Note that surfaces with the best mirror qualities (highest optical reflectivities) tend to be the *poorest* heat radiators.

TABLE 2-4.   Typical Emissivities of Common Surface

Surface	Emissivity, ϵ
Aluminum, Anodized	0.7 — 0.9
Alodine on Aluminum	0.15
Aluminum, Polished	0.05
Copper, Polished	0.07
Copper, Oxidized	0.70
Rolled Sheet Steel	0.66
Air Drying Enamel (any color)	0.85 — 0.91
Oil Paints (any color)	0.92 — 0.96
Varnish	0.89 — 0.93

Courtesy Motorola Semiconductor Products, Inc.

Heat removal by radiation requires that the substances expected to absorb the heat be at a lower temperature than the radiating body. Such absorbed heat would ultimately bring the heat absorbers to the same temperature as the radiator if it were not for the fact that the absorbers discharge their acquired heat to a larger environment external to the radiating and absorbing elements. In actual practice the radiating body tends to lose heat until some equilibrium condition is attained in which both radiator and absorbers arrive at a constant temperature.

We find in physics textbooks that a hot radiating body propagates its heat energy into space at a rate proportional to the fourth power of its absolute temperature. Although it would appear that nature is on our side, there nonetheless remain pitfalls for the unwary. For example, one should not ignore the fact that the same heat-sink surface selected for its high emissivity is also an excellent *absorber* of radiant heat energy. Thus, a transformer with a hotter surface temperature than that of a nearby heat sink will radiate heat *to* the heat sink. Such a situation will prevail even though the intervening air remains cooler than either the *radiator* (the hot transformer) or the *absorber* (the transistor heat sink) because of onvection. Radiation can be employed to lower the temperature of "hot spots" *within* a package, but direct radiation into the *external* environment is usually a more effective use of this thermal technique.

Among the common metals copper is the best heat conductor, aluminum is about half as good, and iron and steel may be six or seven times worse than copper. Brass and bronze, although copper alloys, are generally less than one-third as effective as copper. And lead, despite its physical density, rates less than one-tenth as good a heat conductor as copper. Interestingly these relative thermal conductivities are sequentially the same as the electrical conductivities of these metals.

A not uncommon concern is the wisdom of selecting aluminum or copper as the best material for conducting away heat; this arises because thermal conductivities are usually designated in terms of unit cross-sectional area. If, however, one makes thermal-conductivity comparisons on the basis of density or weight, aluminum emerges as a better material than copper, the same being true of electrical conductivity. Thus, the choice can depend on the requirements of the application. If volume and space merit prime consideration, copper is best. If, on the other hand, as in some airborne projects, the main focus is on weight, aluminum could win. Of course, there are other considerations too, such as cost and the practical difficulty that might be encountered in advantageously using the extra aluminum area interposed in the thermal path. Otherwise both metals are easy enough to machine, cast, and fabricate into various shapes.

Within solid-state power devices beryllium oxide (beryllia) is increasingly being used to conduct heat away from semiconductor junctions to the device case. This material is also encountered as insulating spacers and washers for mounting powr devices on heat sinks. Not only is this ceramic material an excellent heat conductor, but also it is an electrical insulator. This is an unusual but obviously useful combination of physical characteristics. *Be exceedingly cautious when working with devices fabricated with this material or with hardware made of it.* Do not abrade or cut it. In its pulverized or powdered form its absorption or ingestion into the body is hazardous. Although some manufacturers advocate disposal by burial, the emphasis is that it should not be carelessly discarded. It should not be incinerated or allowed to fall into the hands of unaware persons. The mere handling of the solid material is not considered toxic, however.

3

Some Practical Aspects of Solid-State Devices

This chapter continues with the objectives to provide an appropriate preparation for the numerous applications presented in the next three chapters of this book. Practical considerations dictate that some of the ensuing discussion will be construed as review of the reader's arsenal of electronic facts. Some of it will be appraised to be non-relevant by the hardware-oriented worker, while the more theory-inclined may experience frustration with treatments that do not pursue cause and effect relationships to their root sources. However, considering the basic insights in these first three chapters, various and useful benefits will accrue for those whose interest is stimulated by power electronics, notwithstanding their differing viewpoints.

DIODE RECTIFIERS

The basic specifications of current-carrying capacity and reverse-voltage rating are of paramount importance for rectifier diodes. However, there remains the extremely important parameter of *frequency response* or *switching speed* that is all too often overlooked. Although the diode rectifier is generally considered a passive element, its essential function is that of a *switch*. This being the case, little concern was justified with regard to performance at 60 hertz, or even in the several-hundred-hertz region. With present-day power equipment utilizing frequencies in the kilohertz or tens-of-kilohertz range, we must take a second look at the cut-and-dried process of diode rectification.

A deterioration in rectification ability and efficiency sets in at higher operating frequencies. However, this is not the consequence of internal capacitance—at least not in the usual sense. That is, the current leakage through an internal capacitance of up to several tens of picofarads should not appreciably degrade rectification of a 30-kHz sine wave. Why then does the performance of some diodes deteriorate in the audio or the ultrasonic frequency range? Such impairment in rectifying performance, as illustrated in Fig. 3-1, manifests itself in internal dissipation and in highest ripple current in the external circuit.

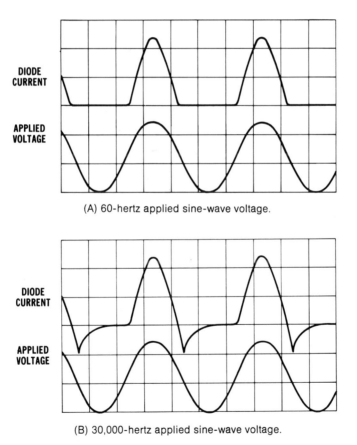

(A) 60-hertz applied sine-wave voltage.

(B) 30,000-hertz applied sine-wave voltage.

Fig. 3-1.   Low- and high-frequency behavior of workhorse junction diode.

Reverse-Recovery Characteristics

It was pointed out in Chapter 2 that the current in a forward-biased junction diode was the net effect of *two* charge carriers, *electrons* and *holes* that are positive charge carriers. It will be recalled that either of these carriers can be made to predominate during the manufacturing process. The predominant, or majority, carrier is due to the impurity atoms in the semiconductor material, and the minority carriers are generated by thermal agitation in the intrinsic atoms of the material. Although the behavior of these mobile charges and that of the fixed ionic charges produces the junction phenomena resulting in rectification, trouble is encountered when too much speed is expected from the minority carriers.

Specifically, if a junction diode is conducting during the forward-bias excursion of an ac voltage, numerous minority carriers participate in the current conduction. But when the ac voltage reverses its polarity, many minority carriers *have not had ample time* to recombine with opposite charges and therefore remain "stored" in both the junction and the bulk semiconductor material. Although these charge carriers are ultimately neutralized or collected, this takes *time*. Therefore, while such charge clearance is occurring, the diode *continues to conduct* despite the reversal of polarity of the impressed voltage. So, for a period of time, called the *reverse-recovery time*, the diode is no longer rectifying. Whether this lapse in rectification is serious depends on how this interval relates to a half-cycle of the applied ac voltage. If the duration of the reverse-recovery time becomes excessive, the diode will heat up from the extra current load that it is forced to carry. Moreover, the rectified dc output to the external circuit will be contaminated with increased ripple current, and filter capacitors may be subject to temperature rise.

Any diode that behaves in this fashion must ultimately experience trouble from low rectification efficiency and internally generated heat if the *frequency of the applied voltage is too high.*

Fig. 3-2 shows four reverse-current recovery characteristics commonly found in junction diodes. The magnitude, as well as the decay path, of the reverse current can be controlled by various processing techniques during manufacture. For such reverse current characteristics to be meaningful when two or more diodes are being compared, the measurements must be made under the *same conditions of forward current and of frequency.* Once this requirement is met, we can determine which characteristic is most desirable.

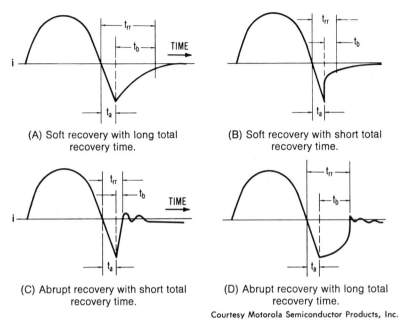

(A) Soft recovery with long total recovery time.

(B) Soft recovery with short total recovery time.

(C) Abrupt recovery with short total recovery time.

(D) Abrupt recovery with long total recovery time.

Courtesy Motorola Semiconductor Products, Inc.

Fig. 3-2.   Various reverse-recovery characteristics in silicon junction diodes.

Other things being equal, the best diode would have no stored charge at all. Then rectification performance out to many tens or hundreds of kilohertz would be substantially as it is in junction diodes at 60 hertz. In practice we must usually choose among such recovery characteristics as shown in Fig. 3-2. Note that the total reverse recovery time is depicted as t_{rr}. This is the time lapse from the initial zero crossing to a current value, which is 25 percent of the maximum reverse current. (Although this is an arbitrary definition, it is a reasonable one.) Also, it is seen that t_{rr} is the sum of *two* intervals, t_a and t_b. The first interval, t_a, is due to charge storage in the depletion region of the junction. The second interval, t_b, comes about by virtue of charge storage in the bulk semiconductor material. The significance of this time resolution is that t_a denotes the time elapsing between zero crossing and the *peak* reverse current. For many practical applications one needs to be concerned primarily with the total recovery time, t_{rr}.

It is also true that a relatively short total recovery time, t_{rr}, may not tell the whole story because one must also know the value of the peak reverse current. In graphical terms it is the area enclosed by the path of the recovery current that is meaningful. In technical terms this area represents the total stored charge (or nearly so). Total stored charge is important because it represents the energy available to produce circuit problems such as ringing, transients, degradation of rectification, and penetration of SOA curves in other devices. Therefore, a vendor should provide either numerical or graphical data relating to total recovered charge. Table 3-1 gives the stored charge in microcoulombs for some popular Motorola diodes. Note that the charge storage for the fast-recovery devices is *significantly less* than for the ordinary diffused diodes. The fast recovery and low charge storage are brought about by a process of gold doping during manufacture. This obviously represents a re-markable improvement in diode performance, particularly at higher frequencies. As may be expected, however, the reduction of charge storage imposes trade-offs in other diode parameters. Specifically the gold-diffused silicon diode tends to have a *lower* reverse blocking voltage and a *higher* forward voltage drop than do ordinary diffused silicon diodes. The relative effects of gold doping on the reverse blocking volt-age are illustrated in Fig. 3-3.

Although "common sense" dictates the use of the rapid recovery characteristic, some power electronic circuits may actually be adversely affected by *too abrupt* a return of the reverse current. This is because of ringing that may be shock-excited by such a rapid rate of change in current. The worst recovery characteristic for nearly all applications is one characterized by high energy storage together with abrupt returns. (High energy storage is usually indicated by high peak reverse current.) If, however, the energy storage is low, the abrupt return is generally favorable. Conversely, if peak reverse current is high, a softer return may be beneficial even though this infers greater energy storage. In

TABLE 3-1.   Stored Charge in Some Popular Motorola Silicon Diodes

Fast Recovery Devices		Diffused Devices	
Type	Typical Charge	Type	Typical Charge
1N4933 Series	0.08 μC*	1N4001 Series	2 μC
MR830 Series	0.12 μC	1N4719 Series	4 μC
1N3889 Series	0.16 μC	MR1120 Series	15 μC
1N3899 Series	0.19 μC	1N1183 Series	20 μC

These values are for a junction temperature of 25°C. The stored charge Q has a positive tem-perature coefficient of approximately 0.25% per degree centigrade.
* μC = microcoulomb

Courtesy Motorola Semiconductor Products, Inc.

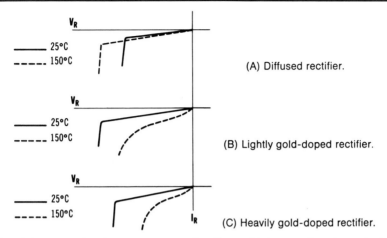

Fig. 3-3.   Example of a parameter trade-off in gold doping of a silicon diode.

many practical applications it proves expedient to strive for low reverse
current *and* a short t_{rr}. This indeed is what gold doping accomplishes.

Other Harmful Effects of Diode Charge Storage

The degradation of rectification is only one of the problems attending
diode charge storage. Inasmuch as rectifier circuits are usually associated
with other solid-state devices, it should come as no surprise that the
functional reliability of such devices may also be impaired. Moreover,
simple rectification is only one function that diodes serve. Other functions include free-wheeling diodes, SCR protection, charging diodes in
LC ringing circuits, circuit-isolating diodes, and commutation diodes.
The charge-storage phenomenon not only can cause faulty operation in
these specific functions but at the same time can adversely affect *other*
devices and their functions.

Such an example is the simple switching circuit shown in Fig. 3-4.
In this circuit the transistor is doomed to failure unless a clamping diode
is connected across the inductive load as shown. Note the collector-
current waveforms in Fig. 3-5. It is clearly seen that the clamping action
of the conventional silicon diode is not nearly as good as that of the
gold-doped, fast-recovery type. Even greater insight into the situation
is conveyed by the load-line diagrams of Fig. 3-6. The entire upper

Fig. 3-4. A simple switching circuit using a clamping diode.

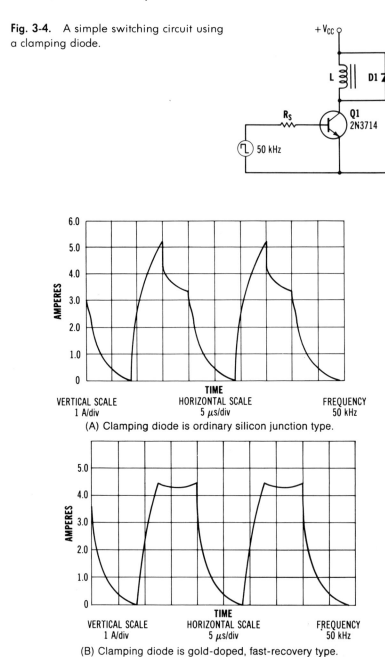

(A) Clamping diode is ordinary silicon junction type.

(B) Clamping diode is gold-doped, fast-recovery type.

Courtesy Motorola Semiconductor Products, Inc.

Fig. 3-5. Collector-current waveforms in switching circuit of Fig. 3-4.

portion of Fig. 3-6A represents needless dissipation in the switching transistor and could penetrate its SOA curve, thereby causing failure. Of course, a lower switching frequency, a trapezoidal switching waveform, or a lower L/R ratio in the load tends to alleviate this problem and often enables satisfactory operation with the use ordinary silicon diodes.

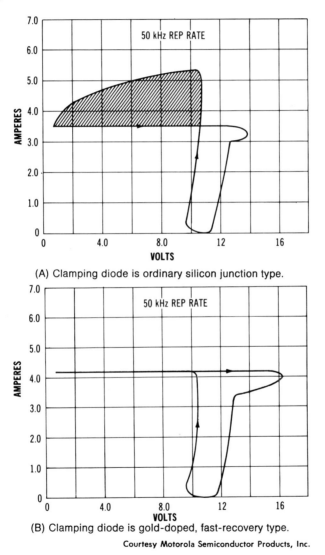

(A) Clamping diode is ordinary silicon junction type.

(B) Clamping diode is gold-doped, fast-recovery type.

Courtesy Motorola Semiconductor Products, Inc.

Fig. 3-6.   Load-line diagram for switching circuit of Fig. 3-4.

Electrical Noise Considerations

Another diode characteristic that merits serious consideration in modern equipment is the rfi and emi potentiality of diodes. One must anticipate many high-frequency harmonics to be present in any source of energy that turns either on or off abruptly. Once these higher-frequency components are excited, it is only natural that some radiation will occur. There is certainly no dearth of "antennae" in the form of printed-circuit boards and interconnecting cables. Often the high-frequency energy backs up into the *power line* from which it radiates only too effectively into other equipment.

It sometimes comes as a surprise that a circuit such as a bridge rectifier can be a vicious source of rfi and emi. This is the result of the recovery characteristic of the diodes and appears to again argue against a too abrupt recovery. However, the rewards of an abrupt recovery are a worthwhile objective in the audio and supersonic frequency ranges. One endeavors to suppress rfi and emi by suitable shielding and filtering techniques and by the use of appropriate components. For example, the use of toroidal-core components is desirable in this regard.

Diodes without Charge Storage

The previous discussion has concerned itself with the manipulation of the reverse recovery characteristic or learning to live with it. Inasmuch as charge storage stems from the presence of minority carriers, the charge storage problem can be eliminated in a diode that accomplishes rectification with *majority* carriers only. Such a device is the *Schottky diode*. It consists of a metal in intimate contact with semiconductor material. With appropriate processing, this does not result in an ohmic contact; neither is a conventional pn junction formed. There is, however, a potential-barrier region at the interface that more or less simulates the behavior of the pn junction. One might view the Schottky barrier region as a half-junction. From the circuit viewpoint two prominent features are provided by the Schottky diode. As mentioned, conduction takes place by majority carriers only and there is virtually no charge storage. Expressed another way, one might say that reverse recovery is instantaneous. (Keep in mind, however, that there is practically *no* reverse current.) The frequency capability of these devices is ultimately degraded by internal capacitance, but this effect does not set in until hundreds of kilohertz or higher.

The second compelling feature of the Schottky diode is its relatively low forward voltage drop. With the increasing demand for low-voltage, high-current dc power supplies, rectification efficiency has become dependent on this parameter. Fig. 3-7 shows a comparison between the forward voltage drops of otherwise similarly rated diodes—one being a conventional pn silicon type and the other being a Schottky type.

We might suspect that the elimination of charge storage in the Schottky diode is possible only at the expense of other parameters. Fortunately the forward voltage drop is not the trade-off, as it tends to be in the gold-doped type of fast-recovery diode. Indeed we do find that the Schottky diode is not particularly good in two of its *reverse* ratings—it tends to have high leakage current and the maximum allowable reverse voltage is about 40 volts or so. This is not a problem when the device is exploited for its low forward voltage drop in low-voltage dc supplies. It does, however, limit its applications. When first available commercially, these devices handled a few tens of amperes and were limited to reverse-voltage ratings of 20 volts, and "junction" temperature was rated at 100°C maximum. At this writing, Schottky diodes are avaiable with current capabilities specified at 75 amperes (continuous), 40-volt reverse-voltage ratings, and at maximum junction temperatures

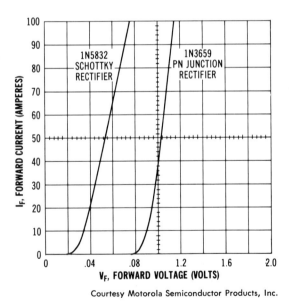

Courtesy Motorola Semiconductor Products, Inc.

Fig. 3-7. Comparison of forward current characteristics of a Schottky diode and a silicon pn junction diode.

up to 150°C. These rectifiers can be used with 20-kHz square waves, whereupon they will deliver clean, almost ideally shaped unidirectional pulses.

The circuit segments illustrated in Fig. 3-8 are frequently encountered in solid-state power electronics. The diodes shown should be either gold-doped, fast-recovery types or Schottky diodes if the operating frequency or switching rate exceeds about 10 kilohertz in most cases. The full-wave, center-tap rectifier of Fig. 3-8D is particularly suitable for Schottky diodes if the output voltage is to be 5 or 10 volts. In this case both the circuit and the diode can contribute maximally to high rectification efficiency.

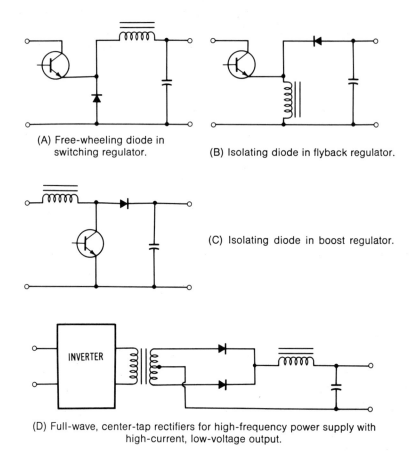

(A) Free-wheeling diode in switching regulator.

(B) Isolating diode in flyback regulator.

(C) Isolating diode in boost regulator.

(D) Full-wave, center-tap rectifiers for high-frequency power supply with high-current, low-voltage output.

Fig. 3-8 Some suitable applications for Schottky diodes and gold-doped, fast-recovery diodes.

THE BEHAVIOR OF SWITCHING TRANSISTORS

An inspection of a large number of power circuits will reveal that the use of transistors as switching devices is at least as important as their classical employment as linear amplifiers or as electronically controlled "rheostats." In theory the process of switching can be 100-percent efficient because the *ideal* switching element exhibits no voltage drop across its terminals when in the on state and involves no current leakage when in the off state. Because a real-life switch such as a transistor is nonideal in both of these respects, it cannot achieve lossless operation. However, there is at least another reason for losses within the switching transistor. Without knowing anything about the inner mechanism of the transistor, we would probably guess that it would not faithfully reproduce a square wave. The simple logic for this is that the vertical sides of a square wave require an infinite number of harmonics of the switching rate, and these harmonics must progress in diminishing amplitude to an infinite frequency. This being so, the most we could expect from a switching device is that it would support a large number of higher-frequency harmonics. But any departure from the ideal harmonic spectrum must necessarily degrade an impressed square wave into a trapezoidal-shaped waveform. The objection to the trapezoid is that the switching device is in its "linear" operating region when tracing out the sloped portions of the trapezoid and therefore incurs dissipative losses in the same manner as any rheostat.

Input and output current waveforms for a switching transistor are shown in Fig. 3-9. Note the "tail" on the input waveform. This is the reverse-current recovery characteristic of the emitter-base diode. During the entire period of reverse recovery, the collector (output) current behaves as if the emitter-base diode continued to be forward biased. This indeed *is* the case, because a diode in reverse-current recovery *is internally* forward biased. (This is another way of describing its charge storage.)

At low switching rates, the presentation of the waveforms in Fig. 3-9 would be nit-picking, for one would obtain an almost-perfect square-wave reproduction, and the input and output currents would be time coincident for all practical purposes. However, when switching at the higher audio and the supersonic rates, the conditions depicted in Fig. 3-9 are no longer of trivial consequence. In addition, there is no guarantee that a 50–50 duty cycle will be reproduced, for it can be seen

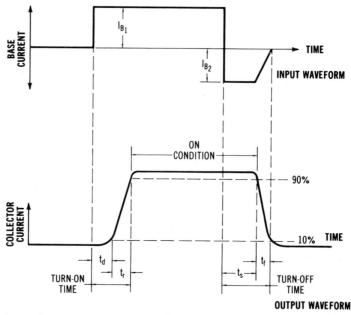

Fig. 3-9.   Input and output current waveforms of a typical switching transistor.

that the duration of the collector current waveform is at the mercy of the charge-storage behavior of the emitter-base region. Indeed, the region t_s is nonexistent if the transistor is not saturated. The length of t_s is proportional to the degree and time the transistor is driven into saturation. For most power control circuits, considerations of operating efficiency dictate a saturation mode with desirably low $V_{CE(sat)}$.

DRIVING POWER TRANSISTORS

A common requisite for all types of power transistors in both linear and switching circuits is proper *drive*. Such a statement might appear so obvious as hardly to merit discussion. However, inspection of such current-gain curves as shown in Fig. 3-10 clearly indicates why circuit designers generally must supplement rigorous design procedures with empirical tests and modifications. Note that the current-gain curves for the Darlington-type power transistor in Fig. 3-10B are plotted on a

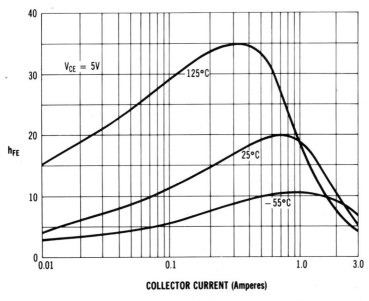

(A) Delco DTS-409 npn tripple-diffused power transistor.

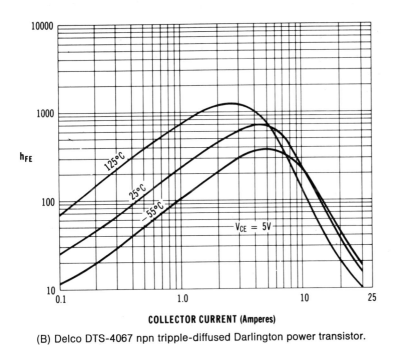

(B) Delco DTS-4067 npn tripple-diffused Darlington power transistor.

Fig. 3-10. Current gain variations in power transistors.

logarithmic scale and represent more extensive variation than does the plot for the ordinary npn transistor in Fig. 3-10A. The simple solution is to provide for *more* drive current than one might initially consider adequate at moderate temperatures and currents.

In switching circuits the first requisite is usually to drive the power transistor well into its collector-saturation region. This is desirable in order to minimize $V_{CE(sat)}$, thereby reducing dissipation during on time. However, driving the base-emitter diode of the transistor *too* hard may prove self-defeating because the input losses then increase. Such emitter-base input losses contribute to the temperature rise of the transistor just as surely as does collector current. Moreover, excessive base current increases charge storage, which can manifest itself as a voltage *spike* in the collector circuit. Such spikes can be destructive because transistor operation is then projected outside the bounds of the SOA curve. Quite often the spikes do not immediately destroy the transistor but seem to be a definite factor in shortening the life of the transistor. At best, these voltage transients add to the electrical noise generated by the switching stage. The extent of this kind of malperformance is not ordinarily predictable due to its dependence on stray circuit inductance.

In linear power control techniques such as amplifiers and dissipative regulators, the consequence of inadequate drive is generally quite obvious because the desired output power will not be available. The competent circuit designer must make adequate drive available under the worst conditions of temperature, line voltage, component aging, and load variation. In feedback systems the phase characteristics of the driving circuitry assume importance, as well as voltage or current capability. Finally, in low-distortion amplifiers it is always a better practice to design the driver for linear response than to use a high amount of loop feedback to compensate for the shortcomings of individual stages because the latter technique often leads to unstable or oscillatory operation.

THE SOA CURVES FOR TRANSISTORS

The term *SOA* is an acronym for *safe operating area*. The development of SOA curves for transistors has been one of the engineering milestones in the field of semiconductor devices. By the intelligent application of such graphical data, it becomes relatively easy to squeeze optimum performance from a power transistor with a sufficient safety factor for reliable operation. Before publication of SOA curves, one more or less flew by the seat of one's pants. There was always a mysterious aura surrounding transistor failure in power equipment, notwithstanding the tabulated specifications on allowable operating conditions. Such specifications are not readily correlated with the *dynamics* of actual transistor operation, even though one might argue that all needed information resides in the numerical limits indicated by the specifications. However, when a load-line diagram is superimposed on the SOA curve, a quick visual inspection substitutes for many laborious calculations and interpolations. (Alternate terms for SOA are *SRO* or *safe region of operation* and *SOC* or *safe operating curves*.)

Consider an ideal transistor for which the manufacturer claims that 50 watts of collector dissipation can be maintained at any combination of collector current and voltage. A simple graph of allowable collector current as a function of applied collector voltage is shown in Fig. 3-11. Mathematically such a curve is described by the equation $E \times I = 50$ (watts). This curve is a true *hyperbola*. An interesting and useful characteristic of the hyperbola is that it becomes a sloping straight line when the plot is made on log-log graph paper. Moreover, the slope is -1, or

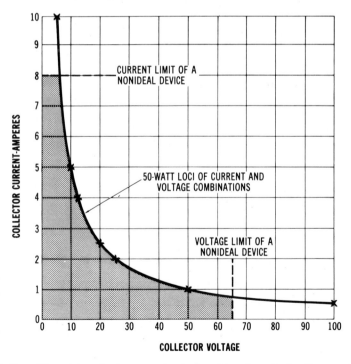

Fig. 3-11. Allowable collector current as a function of applied collector voltage.

45 degrees, no matter what numerical values of voltage and current generate the "curve," and no matter what the log-log scales are. This is shown in Fig. 3-12. The log-log plot has both graphical and numerical advantages over the curve plotted on ordinary coordinate graph paper. The log-log graph accommodates a greater voltage and current range, and better graphical accuracy is obtained because the eye is quick to detect an erroneously plotted straight line. A curve could be quite deviant and yet not be obvious from a quick visual inspection. Also, if the curved plot is made to a different scale, its apparent shape changes, whereas the log-log plot remains a straight line with a 45-degree slope. One or two points suffice to plot an accurate log-log plot, while many points are needed for the curved plot.

Even the ideal transistor cannot be operated at infinite combinations of collector voltage and current. When collector voltage is made

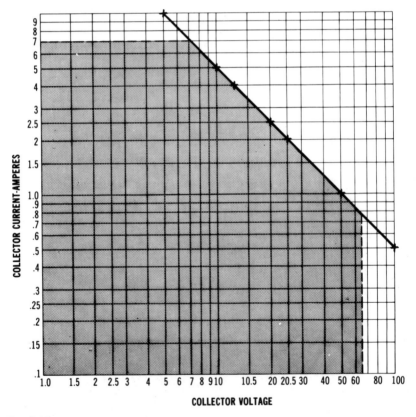

Fig. 3-12. Constant power hyperbola in Fig. 3-11 plotted on log-log graph paper.

too high, avalanche breakdown occurs in the collector-base junction because of the excessive reverse voltage applied across it. Collector current is limited by the minority charge injection from the emitter-base section, which itself is dissipation limited. The graph in Fig. 3-11 clearly shows these allowable limits of operation.

TRANSISTOR BREAKDOWN CHARACTERISTICS

When power transistors were first employed for switches, inverters, motor controllers, and the like, there were many mysterious failures in the field. This was support for the contentions of the die-hards who maintained that semiconductor devices were "naturally" destined to prove unreliable. The failure mode was dubbed *secondary breakdown,* for it was found to resemble and often (but not necessarily) follow avalanche or *first breakdown.* A significant difference was that in most circuit applications the so-called secondary breakdown was much more likely to destroy the transistor. For a while little progress was made in achieving understanding of the destructive phenomenon. The problem could not qualify for solution until a straightforward and reliable rating technique was found that would enable circuit designers to apply it with justifiable confidence. This turned out to be an additional "off-limits" boundary line added to the graph of Fig. 3-12. The resultant graph then became popularly known as the SOA (safe operating area) curve. Fig. 3-13 depicts an actual example of this stratagem. It is seen that there are now four limits imposed on voltage, current, or certain *combinations of the two.* The added restriction of the SOA curve protects against occurrence of the once-mysterious secondary breakdown. In essence, SOA curves inform us that a transistor cannot operate at maximum current, maximum voltage, and maximum power at the same time.

The Two Types of Secondary Breakdown

With regard to secondary breakdown, it turns out that there are *two* distinguishable types. This is not splitting hairs, for one type of secondary breakdown differs from the other in the important aspect that less energy suffices to trigger it. The two types of breakdowns are known as *forward-biased* secondary breakdown and *reverse-biased* secondary breakdown. The biasing refers to the emitter-base section of the transistor. Forward-biased secondary breakdown can occur while the transistor is operating

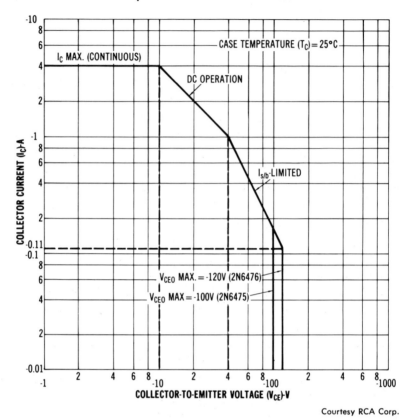

Courtesy RCA Corp.

Fig. 3-13. SOA graphs for two similar RCA pnp transistors.

in its "active" region; that is, collector current flows because of the emitter-base bias. Circuits that come to mind as examples are Class A amplifiers and series-pass regulators.

In contrast, reverse-biased secondary breakdown can occur when the forward emitter-base bias has been removed (or reversed) and the collector current has thereby been switched off. Unfortunately inductance in the collector circuit attempts to maintain the flow of collector current, which leads to the possibility of reverse-biased secondary breakdown. This tends to occur with lower energy levels than does forward-biased secondary breakdown. Contrary to the numerical implications, either breakdown can occur without the transistor having to first undergo avalanche breakdown. Examples of circuits vulnerable to reverse-biased secondary breakdown are inverters and switching regulators.

Differences of Secondary Breakdown
and Avalanche Breakdown

Both types of secondary breakdown are thermally related in that they can be thought of as extremely fast thermal-runaway situations. Secondary breakdowns are caused by severe current concentrations occurring in small semiconductor areas that constitute *hot spots*. That is why secondary breakdown can happen even though the *average* transistor junction temperature is well within specified limits. Such hot spots inject more charge carriers into the collector region, thereby increasing collector current. This in turn adds fuel to the fire, causing even greater temperature rise to be developed in the hot spots. The process is *regenerative* and initially involves such a small mass that the transistor can be destroyed before the average junction or case temperature undergoes appreciable rise.

Thus, secondary breakdown is a *thermal phenomenon*, whereas avalanche breakdown from excessive collector voltage is initiated by a high electric-field gradient. This has important practical ramifications even though the end result of both breakdown phenomena is high collector current. The final limiting voltage following the onset of secondary breakdown is generally in the 5- to 30-volt range. Avalanche breakdown voltage tends to be appreciably higher than this. Secondary breakdown occurs at certain *combinations of voltage, current,* and *time*. The fact that *time* is involved shows that secondary breakdown is basically *energy dependent*. It might appear that with *three* governing parameters, rather than just voltage and current, our ability to guard against secondary breakdown would be destined to remain a black art. That this is not so is clearly revealed by the set of SPA curves shown in Fig. 3-14. (This is actually more typical of published data than is the basic graph of Fig. 3-13.)

We see that penetration of the dc operation boundaries is permissible for *short* on times. The shorter the on time, the more the applicable SOA boundaries extend into *higher* voltage and current areas. This reinforces the previous statement that secondary breakdown is *energy* dependent. Indeed it is true that for *very short* on times the secondary breakdown limit disappears from the SOA curves.

Interestingly, secondary breakdown can be made to take place without necessarily destroying the transistor. Transistor test sets used by manufacturers interrupt the power supplied to the transistor so quickly

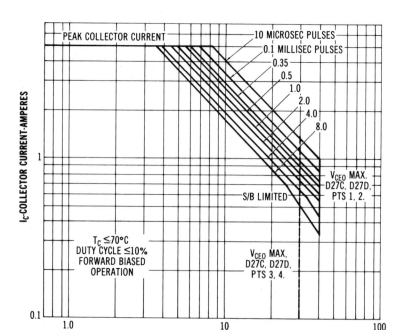

Fig. 3-14. SOA graph for General Electric D27C and D27D power transistors.

at the onset of secondary breakdown that no degradation to the transistor occurs. Designers often build such test equipment to correlate their results better with the true transistor characteristics. However, this is not generally necessary. The quantitative accuracy of published SOA curves is quite good and may be considered reliable for most engineering purposes. Admittedly some skill must be cultivated in order to translate graphical data effectively to practical hardware. Deviations from true standardization are all too common among manufacturers, which does not make the task easier. All things considered, however, SOA curves provide reasonable predictability for the operation of power transistors. Bear in mind that most SOA graphs pertain to forward-bias operation. If an inductive circuit is switched off, one should press the vendor for SOA data in the reverse-bias operating mode because the allowable energy absorption may be appreciably less than for the forward-bias operation.

Temperature Derating

The SOA curves shown in Fig. 3-14 are valid for transistor case temperatures up to 70°C. In practice this usually means that if a suitable
heat sink is used, no temperature derating of the SOA curves is necessary. Often, however, the SOA curves are given for a case temperature
of 25°C. It is not likely that one can use a large enough heat sink to
maintain the transistor case temperature at 25°C, which is only slightly
above room temperature. In order to adapt to practical operating situations, such 25°C, SOA graphs are accompanied by temperature derating curves as illustrated in Fig. 3-15. Before the SOA graphs can be
used, the dissipation-limited boundary lines must be lowered to comply
with the actual case temperature. This is also true for the $I_{S/b}$ line. The
case temperature may be derived from calculations of the thermal circuit

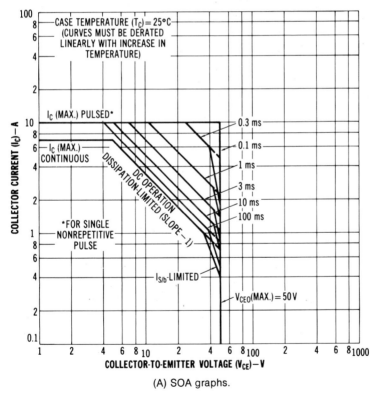

(A) SOA graphs.

Fig. 3-15. 25°C SOA graphs and accompanying derating chart.

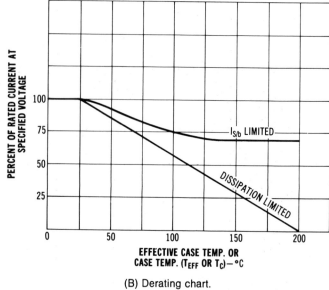

(B) Derating chart.

Courtesy RCA Corp.

Fig. 3-15. *continued*

or may be estimated. Experience plus willingness to experiment somewhat with thermal hardware is usually necessary whether or not thermal
calculations are made, for often variables associated with the packaging
situation are not easy to measure or evaluate. The choice of a higher-
than-actual case temperature is a conservative approach, which allows
some safety factor.

THERMAL CYCLING OF POWER TRANSISTORS

Although there are no filaments to cease functioning, vacuums to become gassy, or contacts to burn and pit, one cannot enjoy the luxury
of ignoring the life span of *power-level* solid-state devices. Along with
lamps, batteries, and electrolytic capacitors, it is often necessary to
ascertain that reasonable service will be obtained, even *after* proper
attention to SOA ratings. (Defiance of SOA ratings is often the cause
of immediate catastrophic failure.) Surely we had hoped that compliance
with SOA ratings and temperature bounds would enable the transistor
to last as long as circuit boards and heat sinks. What indeed "wears

out" in an operating transistor not subjected to transients or other electrical abuse?

It turns out that the life-depleating agent in power transistors is *thermal fatigue*. Somewhere or sometime we have probably encountered this term but because it has often had vague or nebulous connotations, the usual reaction to it has been that it has little place in the listing of predictable phenomena. For example, one might ascribe the unexpected blowout of a fairly new automobile tire to thermal fatigue. Similarly a burned-out light bulb can be conveniently given such an autopsy. Because the life spans of tires and lamp filaments are quite variable, our accounts of their demise are not always convincing. Surprisingly, however, the proper consideration of thermal-fatigue data as applied to *power transistors* can project a reasonably accurate limit to anticipated life span.

Thermal fatigue occurs in power transistors because of differential expansions and contractions of the silicon pellet, the case, solder or weldments, and the bonds used in the construction and packaging of the device. The manufacturer can do the utmost to minimize these thermomechanical displacements but cannot eliminate them entirely. It has been found that the disruption of transistor operation by thermal fatigue is a predictable function of the *number* of thermal cycles to which the transistor is subjected and the temperature excurion of these thermal cycles. Fig. 3-16 shows the thermal-cycle rating chart for the RCA RCS258 npn power transistor. Obviously the circuit designer should strive for effective heat removal and, if feasible, should minimize the number of anticipated temperature cycles.

SELECTION OF TRANSISTOR TYPES

It is quite easy to become confused in the selection and evaluation of power transistors. Not only is there a bewildering array of types, classifications, and fabrications, but the performance characteristics allegedly innate to a particular kind of transistor may vary from manufacturer to manufacturer. A classical example is the long-popular workhorse power transistor, the 2N3055. This device can lay claim to no thoroughbred certification, because its construction, geometry, and chip size have literally varied all over the map. Also it is *not* common for manufacturers' technical literature to play down the current and voltage

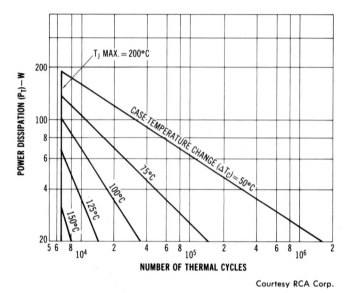

Courtesy RCA Corp.

Fig. 3-16.   Thermal-cycle rating chart for the RCA RCS258 npn power transistor.

capabilities of their products, even when the apparently superb performance pertains to unrealistic situations in practice.

Fortunately, basic guidelines for the selection of transistors do exist. The fact that considerable overlap prevails is an asset rather than a liability to the designer's needs. The generalized listing of power transistor applications presented in Table 3-2 can, at the least, help exclude types likely to be unsuitable. In this table a *low-power* transistor is one capable of dissipating 1 to 10 watts, a medium-power transistor is specified for 10 to 100 watts of dissipation, and 100 watts or more of dissipation qualifies a device as a high-power transistor. If these and the classifications to follow appear arbitrary, they are nonetheless valuable from a relative viewpoint.

Current capability, like power dissipation, is often a ploy of the specification sheets because no rating is absurd if the necessary heat removal can be practically or economically implemented. With this in mind, it becomes useful to consider power transistors with less than 3 amperes of continuous load capability as low-current devices. Medium-current transistors are those capable of continuously supplying load currents from 3 to 10 amperes. And high-current transistors are those with continuous current capabilities exceeding 10 amperes.

Table 3-2. Generalized Applications of Power Transistors

Type	Features	Limitations	Applications
Germanium PNP Former Delco and Motorola types now made by Germanium Power Devices Corporation Andover, Mass.	Low cost. High current capability. Low collector-emitter saturation voltage.	Limited frequency capability. Effective heat removal needed. Collector voltage rating is generally below 100 volts. No npn.	Series-pass elements in linear regulators. Inverters, converters, and switching regulators operating at low and medium frequencies. Audio amplifiers, motor control.
Silicon Single Diffusion NPN	Electrically rugged. Superior thermal characteristics often make this device a better choice than germanium.	Limited frequency range and moderate voltage ratings. Not commonly available in pnp.	Linear regulators and medium-frequency switching applications. Audio amplifiers, motor control, solenoid drivers.
Silicon Epitaxial Base NPN/PNP	Rugged with good frequency range and voltage ratings. Useful for many power-control applications.	Possible marginal performance in high-frequency switching circuits.	General-purpose power device. Audio amplifiers, auto ignition, solenoid drivers.
Silicon Double Diffusion NPN/PNP	Best bipolar device for frequency range and high switching speed. Good current rating.	Ruggedness suffers at higher frequencies. SOA tends to be low.	High-speed switching applications, often used with protective techniques. Rf amplifiers.
Silicon Triple Diffusion NPN	Second-best bipolar device for frequency range and switching speed. Best voltage rating.	Current rating generally lower than other silicon transistors. Cost tends to exceed other "workhorse" types.	High-voltage regulators, inverters, and converters. High-voltage and high-speed switching applications. Rf amplifiers.

Table 3-2. *continued*

Type	Features	Limitations	Applications
Silicon Darlington NPN/PNP	High current gain and high input impedance. Favorable cost and production characteristics.	Trade-offs in frequency, voltage, and current capabilities must be carefully evaluated. Limited rf capability.	Series-pass elements, drivers, moderate-speed switchers, audio amplifiers, auto ignition, motor control, general-purpose power control.
Power MOSFET N-Channel P-Channel	No thermal runaway or secondary breakdown. Very high switching speed and input impedance. Parallels easily.	"On" resistance may cause excessive voltage loss. Vulnerable to gate damage during handling.	Linear and switching-type power supplies. Audio amplifiers, rf amplifiers, logic-driven power control. Increasingly competitive with bipolars.
Synthesized Transistor	Overload protected. Base drive as high as 40 volts without damage. 0.5-μs switching time. 3-μA base current. Interfaces CMOS or TTL.	Limited current and voltage ratings. (Approx. 2 A and 40 volts) No pnp version presently available. Vulnerable to self-oscillation.	High-reliability, low-power output element in both switching and linear supplies. General-purpose power control.
Hybrid circuit Module NPN Output	Overload protection. Higher power than monolithic devices. Op-amp circuit characteristics. Electrically rugged.	More costly to make than monolithic ICs. Moderate frequency capability. Power capability less than simple transistors.	Regulated power supplies. Public address systems. Motor control. Solenoid drivers. Deflection circuits. Servo amplifiers. Driven inverters. Workhorse for 5- to 100-watt loads.

Voltage ratings are sometimes harder to pin down because of recent progress in this area. However, because of the predominance of silicon over the germanium technology, it appears reasonable to consider any power transistor with a voltage rating of less than 90 volts as a *low-voltage* type. The *medium-voltage* classification would then embrace the range of 90 to 300 volts. Beyond this range are the *high-voltage* transistors that were almost inconceivable not too long ago but now commonly used in tv sets and for other applications.

In the frequency or speed domain, power transistors capable of switching up to 3 kilohertz without incurring appreciable rise-time and fall-time losses can be considered to be low-frequency devices. From 3 to 15 kilohertz comprises the middle-frequency range, and a power transistor capable of operating efficiently at higher switching rates qualifies as a high-frequency type. This generally implies that the cutoff frequency, F_T, will be well into the megahertz range. (The cutoff frequency of some older, workhorse germanium transistors, such as the 2N178, was a mere 10 kilohertz.)

Another helpful guide in transistor selection is the realization that the operating parameters have a strong tendency to be interrelated on a *trade-off basis*. One can buy germanium power transistors capable of withstanding 500 volts, but their already poor frequency ratings are further degraded and their low-cost feature is lost. There are two subdivisions of the triple-diffused silicon transistor, one optimized for frequency at the expense of voltage, and the other with high voltage and lower frequency ratings. In Darlington transistors the high current gain is attained with a sacrifice in speed of response. (It is indeed noteworthy that modern monolithic Darlingtons have been successfully competing with other types in a wide variety of power-control applications.)

Because of the diversity of transistor types and the large number of processing and construction techniques, it should be appreciated that no classifying system can be truly comprehensive. Indeed, if such a system existed, its complexity would defeat its objective. Table 3-2 endeavors to generalize power-transistor use in terms of what now appears to be significant structural differences. An example of a type not listed in the table is the RCA pi-nu multiple epitaxial-base transistor. This transistor structure appears to have blended successfully parameters otherwise tending toward *either* the double-diffusion or ordinary epitaxial-base types. As a consequence, we find this power transistor enjoying widespread application in original-equipment auto ignition circuitry.

The RCA 2N6513 exemplifies the ratings of pi-nu devices. Its current rating is 5 amperes and it can withstand 350 volts. Its F_T specification is 3 megahertz and maximum power dissipation is 120 watts. It can readily be appreciated that such a parameter combination optimizes this transistor for service in solid-state ignition systems.

Various transistor structures are shown in Fig. 3-17. Unfortunately these illustrations are not likely to be as meaningful as the more visible features in vacuum-tube structures. Nonetheless it is instructive to contemplate the different processing techniques that are responsible for the diversity in the operating parameters of transistors.

Planar construction tends to produce devices with less leakage current than the earlier developed *mesa* process. In a planar transistor the junction edges are encapsulated in silicon dioxide "glass." This also tends to reduce noise generation (admittedly, not generally important in power devices) and enhance reliability. *Epitaxial* structures are those in which the manipulation of impurity regions in the semiconductor material is achieved by induced crystalline formation. The thickness and electrical sensitivity of such layers of "grown" material can be precisely controlled. Epitaxial fabrication techniques are particularly useful in lowering collector saturation resistance and in increasing the voltage capability of transistors.

The Darlington transistors have been improved to the extent that they are competitive to the various single-transistor types in a wide range of applications. Where it can be substituted for conventional transistor types, the Darlington transistor often brings about both operating and economic advantages. Because of its high current gain, it can often be directly driven from integrated circuits. When comparing the parameters of all of the power transistors listed in Table 3-2, one finds that the Darlington appears to have the best mix of medium ratings. At the same time the Darlington lends itself well to optimization of certain parameters. An example is the line of high-voltage Darlingtons available from Delco Electronics. The Delco DTS4075 is a monolithic Darlington with built-in resistors between each base and emitter, as well as an internal speed-up diode to improve switching characteristics. This device has a maximum collector-current rating of 20 amperes. It withstands 350 volts at 1 ampere of collector current.

The synthesized transistor, as exemplified by the National Semiconductor LM195, is a specialized IC. This power device is both inexpensive and cost-effective because it incorporates its own reliable protection

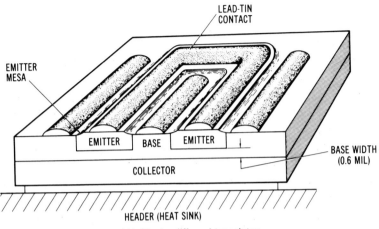

(A) Single-diffused transistor.

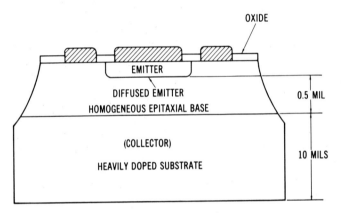

(B) Epitaxial-base transistor.

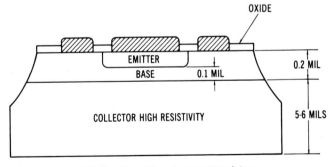

(C) Double-diffused mesa transistor.

Fig. 3-17. Some examples of silicon transistor structures.

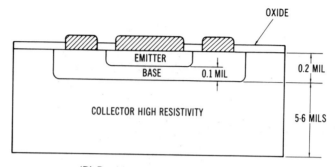

(D) Double-diffused planar transistor.

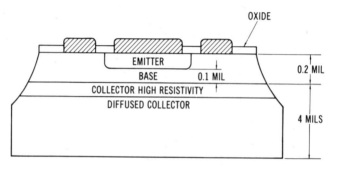

(E) Triple-diffused transistor.

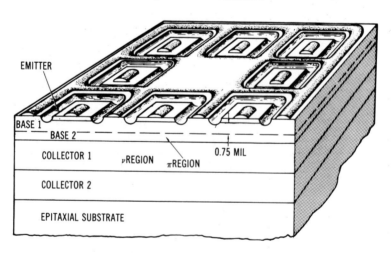

(F) RCA pi-nu multiple-epitaxial-base transistor.

Courtesy RCA Corp.

Fig. 3-17.   *continued*

against current overload or temperature rise. Therefore, one can dispense with external protective devices and circuitry, as well as the time-consuming experimentation often needed to confirm the integrity of protection techniques. Although a relatively low-power device, it is nonetheless a far cry from earlier attempts at IC power devices. At this time the company is bringing out a "pnp" version. Despite its low cost, the LM195 is certainly among the most sophisticated of the low- and medium-frequency power-control devices, for its monolithic architecture comprises more than 20 equivalent transistors.

While on the subject of nondiscrete power devices, it is appropriate to mention thick-film assemblies and hybrid modules. Although not of monolithic fabrication, such "power packages" resemble the synthesized transistor in the makeup of their internal circuitry—several active devices are connected so as to provide an easy-to-use power-control device. An example is the RCA HC2000H power op-amp. This thick-film–hybrid module is, as its name implies, an operational amplifier with differential input. It can supply peak output currents as high as 7 amperes and can deliver rms power up to 100 watts. Its bandwidth of 30 kilohertz enables it to render useful service in audio, servo, and deflection amplifiers and in voltage regulators and driven inverters. Such modules seemingly combine some of the best features of discrete solid-state devices and monolithic ICs.

Double- and triple-diffused power transistors merit consideration when either high frequency or high voltage imposes the need for uniquely optimized devices. These structures are not noteworthy for their current-carrying ability or a high SOA. The specialty of double diffusion is speed, while that of triple diffusion is high voltage rating. Depending on doping, geometry, and other aspects of the manufacturer's processing techniques, there can be considerable overlap in application areas for the two types. Also, when using these types, it should be kept in mind that although both switching and rf transistors have the need for speedy response, best results are obtained by ordering devices that are *uniquely specified* for one or the other service. In switching transistors, for example, one is more concerned with a low collector-saturation voltage than in rf types. Rf transistors are required to endure unusual stresses resulting from maladjusted tuning and from high vswr. Somewhat different considerations apply to the internal capacitances of the two types, although these capacitances must be low in both instances. In addition, rf types are generally designed for lower base-emitter voltage but greater input current than the switching types. Because of the widespread use of mobile communications, rf transistors are more often intended for 12- or 24-volt operation, whereas switching types may operate at considerably higher voltages. Other subtle differences become more pronounced at high frequencies and emphasize the need to specify the intended service.

Silicon single-diffusion and epitaxial-base transistors are more appropriate for the numerous power-control functions where undue emphasis is not placed on speed or voltage capability. Today's workhorse devices have parameters that were the domain of the sophisticated devices of yesterday. Accordingly both of these structures find application in inverters and switching circuits where good 20- and 30-kilohertz square waves must be produced. Here some of the older, general-purpose germanium transistors would not have been useful.

At lower speeds the germanium transistor still has its devotees. If one is prepared to apply more effective heat removal than is required from silicon devices, use of the *germanium* transistor can often prove satisfactory. It even has the compelling feature of exceptionally low collector-saturation voltage. So, far from being obsolescent, the germanium transistor often finds use in modern power supplies that operate at low voltage and high current. It is still being used in automobile radios and in economy audio equipment. Although fewer companies make germanium power transistors, a residual demand remains. Because of their high current ratings, tolerance to overload, and low cost, many designers still feel that germanium transistors are entirely justified for certain types of motor control, regulators, solenoid drivers, etc.

BASIC TYPES OF SCRs

Many improvements have been incorporated into SCRs since they first became commercially available. These improvements have been targeted at such parameters as triggering reliability, commutation performance, noise immunity, ability to endure and perform in the face of rapidly changing voltage and current, speed, and, of course, voltage and current capability. A major dividing line for SCRs pertains to *speed*. The SCRs that are intended for *60-hertz phase-control* service are grouped together, as are those intended for *inverter* service. *Phase-control* SCRs often work well at 400 hertz, but at higher frequencies the *inverter* SCR yields optimum performance. Many inverter SCRs appear to be specified for frequencies in the 3- to 10-kilohertz range, but recent developments have pushed the top frequency to the 20- to 30-kilohertz range and some are available with even higher frequency capability. The terms are somewhat misleading, for it is not the phase control or the inverter operation per se that is significant. Rather it is the frequency at which the SCR must turn on and off. It just happens that most 60-hertz applications utilize phase control, and the higher frequencies are often associated with inverters.

As with transistors there are different techniques for SCRs. The low-cost workhorse for many practical applications has been the *alloyed-diffused* SCR. These SCRs have been suitable for 60-hertz control applications up to 35 amperes and are available with blocking voltage ratings up to 800 volts. These numbers are not intended as rigid boundaries, but when audio frequency rates or greater current or voltage are involved, SCRs constructed via *epitaxial* technology generally merit attention. The SCRs for even more demanding service, where perhaps 2,000 volts or more must be blocked together with control of many hundreds of amperes of load current or with frequencies of 5 kilohertz or higher, require the most sophisticated of all structures. These devices may combine epitaxial processes, the use of a shunt between gate and cathode, and gold doping. Invariably they contain special gate structures designed to propagate the turn-on action rapidly, as there tends to be a time lapse between the trigger pulse and the anode-cathode response. Large SCRs also must be equipped with effective *thermal hardware.*

APPLICATION OF SCRs AND TRIACS

The most important members of the thyristor family of solid-state power-control devices are SCRs and triacs. They offer efficient control of high power levels and generally involve relatively simple circuits. Nonetheless successful implementation can be achieved in many instances only by paying attention to a number of important characteristics of these devices. Although the semiconductor manufacturers endeavor to make applications as simple and straightforward as possible, nothing substitutes for knowledge on the part of the user in the quest for trouble-free performance.

In the following discussion the SCR will be referred to except where it is desired to deal with the unique features of the triac. This is possible because many of the important operating mechanisms are shared by the two devices. As yet, however, triacs lag behind SCRs in voltage, current, and frequency ratings.

Forward Voltage Drop

A commonly encountered notion is that the voltage drop across the anode-cathode terminals of an SCR operating in its conductive state is at least two volts. The reasoning involved is that there are *three* pn junctions in series with a 0.6- or 0.7-volt drop across each one. In actual devices the measured voltage drop is only that attributable to a single junction plus another few tenths of a volt. For low to medium currents, this voltage drop is in the neighborhood of 0.8 to 1.3 volts, depending on certain variables such as the device fabrication and the amount of current being handled. The discrepancy stems from the erroneous viewpoint of the device as being *equivalent* to three series-connected diodes. A little reflection will reveal the invalidity of this analogy inasmuch as the "middle diode" is not properly polarized for forward conduction.

This dilemma is resolved by recalling the two-transistor simulated SCR of Fig. 2-24 in Chapter 2. The fact of the matter is that, although the SCR is not *constructed* as a pair of separate transistors, it does indeed *function* as a two-transistor device. The principal current path is through the emitter-base section of a pnp "transistor" and the collector-emitter circuit of an npn "transistor." The regenerative action of the "transistor" pair holds both "transistors" in heavy saturation. The emitter-base section of the pnp "transistor" can behave only as an ordinary diode, contributing about 0.6- to 0.8-volt drop. However, an entire transistor in saturation, in this case the npn "transistor," can have a collector-emitter voltage drop *appreciably less* than that of a diode alone. Thus, the *net* voltage drop from anode to cathode of the SCR is only in the vicinity of that corresponding to some rectifier diodes that have not been processed for an optimally low voltage drop.

Triggering of Thyristors

The most common thyristors used in power electronics are SCRs and triacs. They share much in common with respect to triggering techniques because proper triggering is one of the prime operational requirements of both types of thyristors. In the ensuing discussion particular reference may be made to SCRs, but unless otherwise stated, similar circumstances will also pertain to triacs.

Gate voltage and gate current for thyristors have relatively wide tolerances from unit to unit. Moreover, the gate trigger requirements are also temperature sensitive and also vary with the amount of voltage at the anode before conduction. A typical example of gate-triggering characteristics for a good quality SCR is shown in Fig. 3-18. One might conclude from these characteristics that a trigger source with plenty of voltage and with adequate current capability should solve any problems stemming from a wide range of gate sensitivities. Although there is considerable truth in such a proposal, it cannot be applied with reckless abandon. The gate is limited in its power-dissipation rating. Also, excess gate power lowers overall efficiency and subtracts from the allowable anode-cathode dissipation. This is particularly true at higher frequencies. Often the amount of reverse voltage impressed on the gate is a limiting factor.

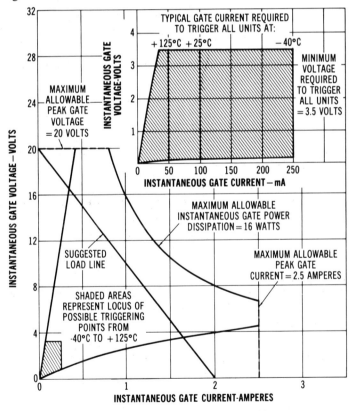

Courtesy General Electric Co.

Fig. 3-18. Example of SCR gating characteristics.

Although adequate gate voltage and current might assure the triggering of all thyristors of the same type, they would not solve the problem of divergent control settings for a given load power in phase-control equipment. Ideally in phase-control applications the triggering of the thyristor should be governed *only* by the time constant or delay imposed by the trigger circuit. The time of triggering should *not* be influenced by the amplitude of the gate signal. Finally, this brute-force approach would not prevent inadvertent triggering of the most sensitive thyristors by *noise*.

Before designing appropriate trigger circuitry to comply with these variables, the width of the trigger pulses should be considered. It is obvious that if the pulses persist too long after the thyristor has made its intended response, gate dissipation is bound to be unnecessarily high. We can even get into timing difficulties if the trigger signal remains when the thyristor is not intended to be conductive. On the other hand the trigger pulse cannot be made infinitesimally *short* in duration, because the triggering phenomenon requires that a threshold quantity of electrical charge by delivered to the gate. The effect of pulse width for a particular SCR is illustrated in Fig. 3-19. This happens to be a high-frequency SCR, but the curves are qualitatively similar for all thyristors. In 60-hertz types the *knee* tends to be in the vicinity of 5 to 10 hertz.

Triggering Devices

A number of solid-state triggering devices are available for thyristors. Table 3-3 lists the most frequently used types. All these triggering devices develop sharp pulses when used in appropriate circuits. The peak voltage, peak current, and pulse duration of these devices are compatible with a large number of commonly used thyristors. With large thyristors the gate current requirement may necessitate the use of an emitter follower or a small intermediate thyristor as a buffer or driver.

The original diac symbol, depicting a transistor structure with two "emitters," was relevant to the first devices marketed for thyristor triggering. This symbol is now often replaced by the symbol for a "gateless thyristor" (four-layer diode). Whether the actual device referred to as a diac is fabricated with three or four layers, the bilateral switching behavior is much the same.

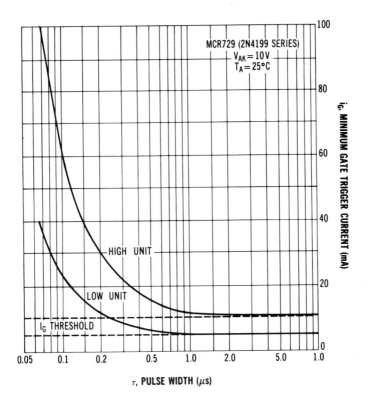

Fig. 3-19.   Effect of trigger pulse width on minimum required triggering current.

COMMON SCR TRIGGERING CIRCUITS

The following circuits illustrate a number of commonly encountered triggering arrangements for SCRs. In the simple circuit of Fig. 3-20 a neon lamp is used as the trigger device. As with many bargain items, this circuit could prove expensive in the long run. One of the disadvantages is the well-known erratic firing characteristics of neon lamps. Aggravating this problem are the unpredictable effects of aging, temperature, light, and electric fields. If this scheme is used, it is recommended that a neon lamp intended for such purpose be selected rather than the garden-variety type.

TABLE 3-3. Solid-State Triggering Devices for Thyristors

CLASS	E-1 CHARACTERISTICS	BASIC CIRCUIT	MAJOR TYPES	V_p PEAK POINT VOLTAGE	I_p (MAX) PEAK POINT CURRENT	I_v (MIN) VALLEY CURRENT	T_{ON} TURN ON TIME	SPECIFICATION NUMBER
UJT Unijunction Transistor 	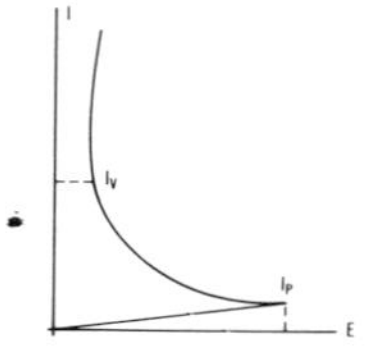	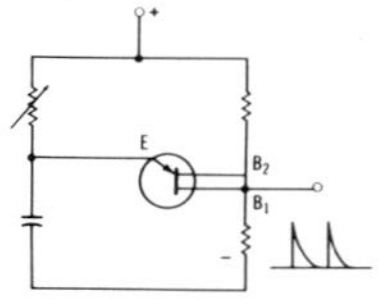	TO-5 TO-18 1N489A 2N2417A 2N489B 2N2417B 5G515 2N1671A 5G516 2N1671B 2N1671C 2N2646 2N2647	Fixed Fraction of Interbase Voltage	12 μA 6 μA 25 μA 6 μA 2 μA 5 μA 2 μA	8 mA 8 mA 8 mA 8 mA 8 mA 4 mA 8 mA	1-2 μsec Typ	60.10 60.10 60.50 60.50 60.02 60.62
CUJT Complimentary Unijunction Transistor 	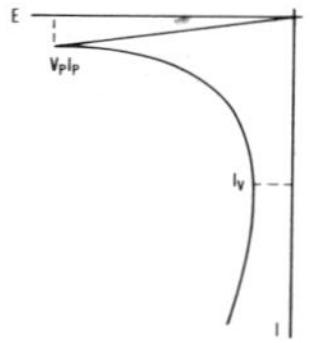	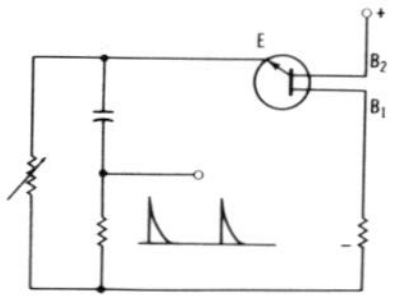	D5K1 D5K2	Fixed Fraction of Interbase Voltage	5 μA 15 μA	1 mA 1 mA	1 μsec Typ	60.15 60.16
PUT Programmable Unijunction Transistor 	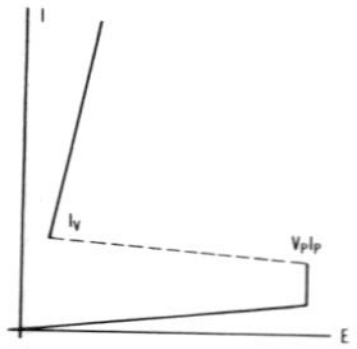	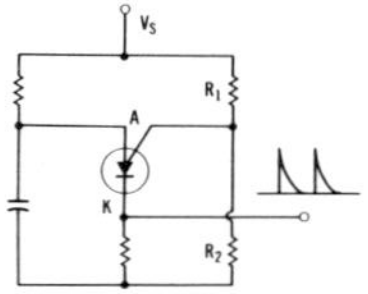	D13T1 D13T2	$\dfrac{R_2 V_s}{R_1 + R_2}$	As low as 2 μA As low as .15 μA (A function of R_1, R_2)	70 μA 25 μA (A function of R_1, R_2)	80 nsec Max	60.20

Unidirectional

144

TABLE 3-3. continued

	Type Numbers	Voltage	(µA)	(mA)	Time	Reference
SUS Silicon Unilateral Switch 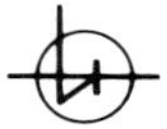 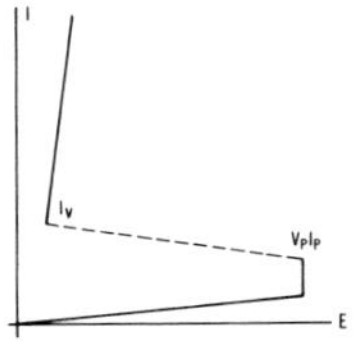 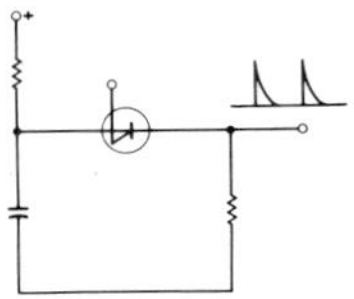	TO-18: 2N4983, 2N4984, 2N4985, 2N4986 — TO-48: 2N4987, 2N4988, 2N4989, 2N4990	6.10 V / 7.5-9 V / 7.5-8.2 V / 7-9 V	500 µA / 150 µA / 300 µA / 300 µA	1.5 mA / .5 mA / 1.0 mA / .75 mA	1.0 µsec Max	65.25, 65.26 / 65.27, 65.28 / 65.27, 65.28 / 65.25, 65.26
SCS Silicon Control Switch 	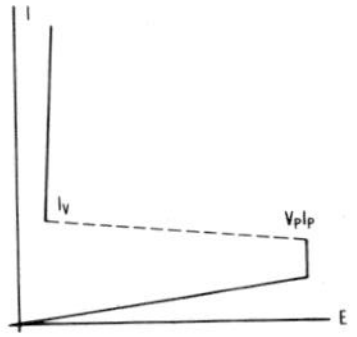 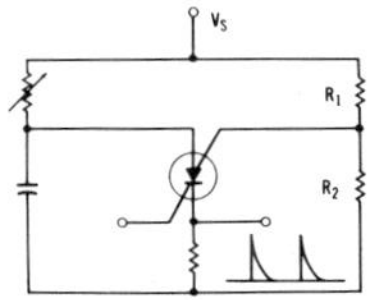 3N84	$\dfrac{R_2 V_s}{R_1 + R_2}$ (40V max)	A function of R_1 and R_2	10 mA Max (A function of R_1 and R_2)	1.5 µsec Max	65.18
SBS Silicon Bilateral Switch  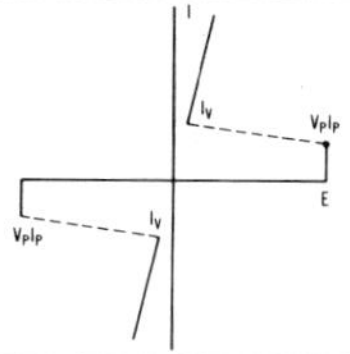 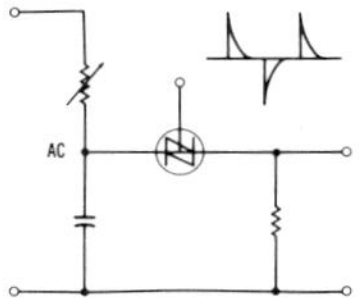	TO-18: 2N4993 — TO-98: 2N4991, 2N4992	6.10 V / 7.5-9 V	500 µA / 120 µA	1.5 mA / .5 mA	1.0 µsec Max	65.30, 65.31 / 65.32
DIAC 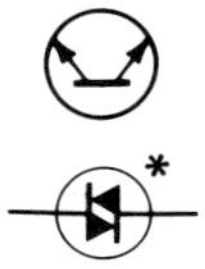	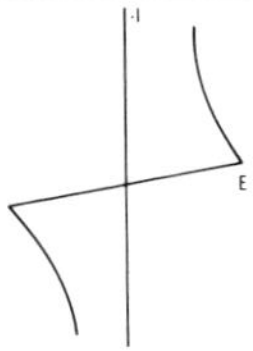 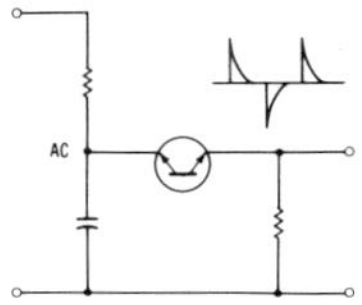 ST2	28V-36V	200 µA	Very high	1 µsec Typ	175.30

Bidirectional (SBS, DIAC)

* Alternate Symbol

Courtesy General Electric Co.

145

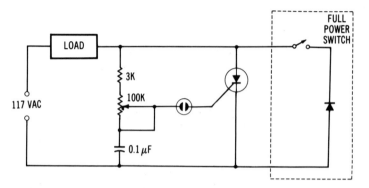

Fig. 3-20. A neon lamp used as a trigger device.

The circuit shown in Fig. 3-21 employs a diac in conjunction with a pulse transformer having a 1:1 turns ratio to obtain full-wave phase control from a pair of back-to-back SCRs. The overall operation simulates that ordinarily obtained with a triac. The circuit in Fig. 3-22 utilizes a unijunction transistor fed from rectified ac. The important thing in this arrangement is that no filter capacitor should be used with the dc supply because it is necessary that the operating voltage be periodically returned to zero.

In the half-wave phase-control circuit of Fig. 3-23 the trigger device is a silicon bilateral switch. This is an IC with the simulated characteristics of the diac. The structure involves neither the four-layer fabrication nor the double-emitter transistor commonly employed in bilateral switches. In this case the circuit function is *synthesized.* Another difference from the diac is the availability of one or more *gates,* which may be optionally used for various logic techniques. In this application the use of the gate always causes the capacitor to commence its charge cycle from the same near-zero voltage. This makes the control hysteresis-free and prevents the irritating "snap-on" effect, which plagues many "economy-type" lamp dimmers. The diode protects the SCR gate from reverse biasing.

A somewhat similar trigger scheme is shown in circuit Fig. 3-24, which uses a silicon unilateral switch. A unique feature of this circuit is that the power can be controlled in either a dc or an ac load. The dc output voltage from the bridge rectifier should not be filtered, however, because the commutation of the SCR is achieved by the periodical return of this voltage to zero.

Fig. 3-21. An SCR triggering circuit using a diac.

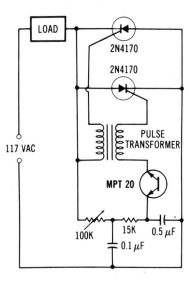

Fig. 3-22. A unijunction transistor is used in this SCR trigger circuit.

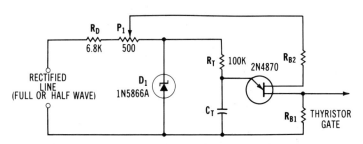

Fig. 3-23. A half-wave phase-control circuit using a silicon bilateral switch as the trigger device.

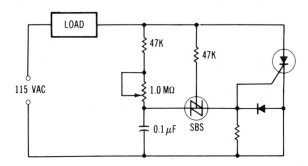

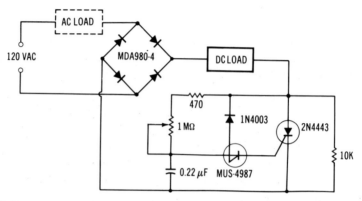

Fig. 3-24. A triggering circuit using a silicon unilateral switch.

The trigger circuit in Fig. 3-25 makes use of an optocoupler. The salient feature of this triggering technique is the electrical isolation obtained between control circuitry and the load. Optocouplers with 1,500-volt isolation between the input LED and the light-sensitive output device are commonly available. Higher voltage units are also marketed. Triggering via optocouplers is especially popular in solid-state relays. This technique also merits consideration in systems where it is necessary to prevent false triggering from noise or transients.

Any mention of optoelectronic techniques naturally suggests the light-actuated SCR. These devices inherently provide electrical isolation between control or trigger circuits and the load. Fig. 3-26 depicts an arrangement of light-actuated SCRs for full-wave control of load power from a high-voltage ac line. All of the gates are illuminated simultaneously. Sometimes this is achieved with the use of fiber optics. As

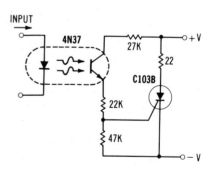

Fig. 3-25. An optocoupler used to trigger an SCR.

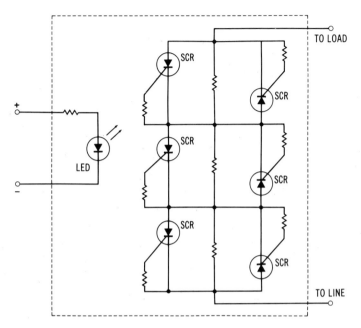

Fig. 3-26.   Full-wave control of load power by light-activated SCRs.

shown, the circuit can be used as a switch by maintaining the gate "signal" as long as it is desired to energize the load. Such triggering does not cause gate dissipation as do all electrical gate-triggering methods.

There is more than meets the eye in the partial circuit of Fig. 3-27 using the General Electric PA424 module, which was one of the first successful zero-crossing voltage detectors. As shown, the circuit delivers full-wave power to the two load SCRs. When this circuit is switched on, it applies load current only when the voltage sine wave is crossing its zero level. Such a scheme generates negligible rfi, and therefore filters are generally not needed. Control of load power is usually accomplished by burst modulation—varying numbers of *complete* ac cycles are applied to the load. Such bursts are interspaced by dead times in which no current is available for the load. The complete circuitry for doing this is not shown because the present focus is on the triggering technique itself. A number of ICs are available for accomplishing zero-voltage control and for facilitating feedback arrangements so that the

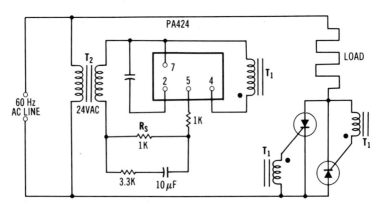

Fig. 3-27. A trigger circuit using a zero-crossing voltage detector.

desired control or regulation can be realized. This triggering method spares thyristors from considerable stress because both *dv/dt* and *di/dt* are much less than is likely to be encountered in phase-control circuits.

The phase-control triggering circuit of Fig. 3-28 uses a programmable unijunction transistor (PUT). A noteworthy feature of this arrangement is that the firing point of the PUT can be adjusted by varying the resistance associated with its anode gate. In essence we have a variable diac.

The triggering circuit of Fig. 3-29 provides essentially the same kind of load power control as the arrangement shown in Fig. 3-21. An advantage of the configuration of Fig. 3-29 is that total electrical isolation is obtained between the load and the trigger circuitry. However, a three-winding pulse transformer is required.

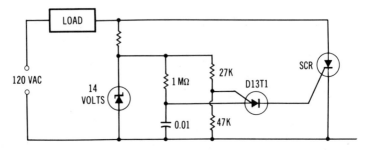

Fig. 3-28. A phase-control triggering circuit using a programmable unijunction transistor.

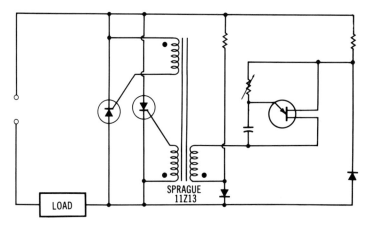

Fig. 3-29. Circuit providing alternate triggering of two SCRs.

The transistor circuit shown in Fig. 3-30 can be tailored to operate in either of two modes. It can perform as a linear emitter follower. Its chief usefulness, however, results when the transistor is driven into heavy saturation. It then acts as a *switch,* which abruptly transfers the charge accumulated in capacitor C1 to the gate circuit of the SCR.

The trigger circuit in Fig. 3-31 is best termed a *pulse sharpener* because it will transform a slow-rising pulse into one having a rise time

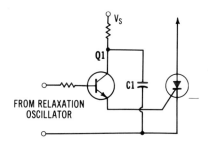

Fig. 3-30. An SCR triggering circuit using a bipolar transistor.

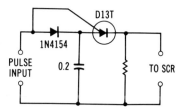

Fig. 3-31. Trigger circuit using a programmable unijunction transistor as a pulse sharpener.

of 50 to 100 nanoseconds. When a relatively slow positive-going pulse is applied to this circuit, the 0.2-microfarad capacitor commences charging through the 1N4154 diode. As long as the charging process continues, the D13T programmable unijunction transistor (PUT) will not fire because the anode gate will be *more* positive than the anode by at least the voltage drop of the diode. However, when the pulse attains its peak value and starts its decline, the anode gate will become *less* positive than the anode because of the retention in the capacitor of the near-peak voltage level. This condition, enhanced by the fact that the diode becomes reverse biased, enables the PUT to fire. As mentioned, the resultant SCR trigger pulse then has an extremely fast rise time.

The arrangement illustrated in Fig. 3-32 has two implementations. The predriver SCR may be external to the main SCR, or it may be integrally fabricated with the main SCR. In the latter instance the composite structure is known as an *amplifying gate SCR*. Such a device often enables the circuit designer to incorporate a high-power SCR into a system with the use of the same trigger circuits ordinarily employed at lower power levels.

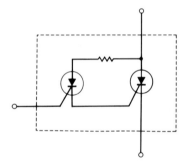

Fig. 3-32. Amplifying gate SCR used to trigger main SCR.

When the SCR has to turn on fast-rising load currents, reliable triggering often demands a rectangular pulse with fast rise time. A better approach to this requirement than any of the previous circuits is the circuit shown in Fig. 3-33. Here a UJT relaxation oscillator is followed by a cascade arrangement of four bipolar transistors that takes advantage of the cumulative storage time to lengthen the pulse. The net result is a fast-rising rectangular trigger pulse about 10 microseconds in duration that has a 20-volt amplitude with a source impedance of 20 ohms. It can be produced at a rate of up to 20 kilohertz.

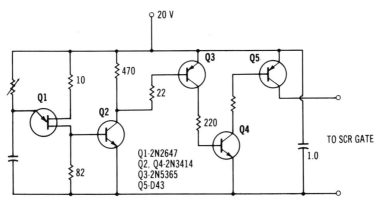

Fig. 3-33. An SCR triggering circuit using a cascade of four bipolar transistors to lengthen pulse.

The triggering circuit shown in Fig. 3-34 provides sustained excitation of the SCR gate but minimizes power dissipation through a sequence of narrow pulses rather than a prolonged rectangular wave. Such sustained triggering is often required when a highly inductive load is being controlled. Another feature of this arrangement is that two pulse trains displaced 180 degrees in low-frequency time are generated. (One

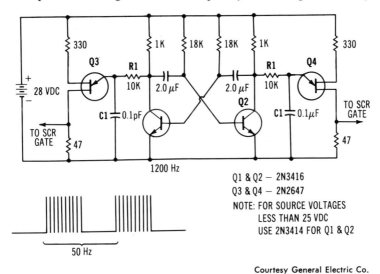

Courtesy General Electric Co.

Fig. 3-34. Trigger circuit providing a sequence of narrow pulses to minimize power dissipation.

sequence of high repetition rate pulses is on while the other is off, and vice versa.) Such an operational mode is ideal for inverters. The alternations are brought about through the action of the multivibrator consisting of transistors Q1 and Q2 and associated circuitry. When transistor Q1 is in its conductive state, the relaxation oscillator including UJT Q3 is prevented from producing the high-repetition-rate trigger pulses. Similar inhibitory action occurs during the subsequent half of the multivibrator cycle when UJT Q4 is prevented from oscillating because of the conductive state of transistor Q2.

A somewhat similar, but simpler, triggering circuit for inverters and push-pull or bridge-type power-control systems is depicted in Fig. 3-35. The trigger pulses occur in flip-flop fashion, enabling one SCR of a pair to be triggered when the other SCR is off. Single sharp pulses are generated rather than a burst. The circuit consists of two PUT relaxation oscillators arranged so that a trigger pulse generated by one of them commutates the other oscillator. In this way each oscillator is allowed to generate only one pulse at a time. Commutation takes place via capacitor C_T. Adjustment of the repetition rate is provided by potentiometer R1. Symmetrical spacing of the pulses can be attained by adjustment potentiometer R2.

A series-connected arrangement of SCRs, which is able to hold off many thousands of volts, can be utilized in high-voltage inverters and switches. It is, however, necessary that all of the gates be triggered

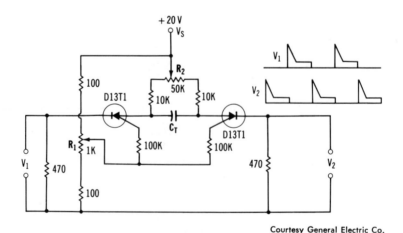

Courtesy General Electric Co.

Fig. 3-35. Triggering circuit for inverters and push-pull or bridge-type power-control systems.

simultaneously. It is also desirable that the gate circuits be electrically isolated from the high-voltage system. A convenient way of meeting these objectives is with a triggering scheme such as in Fig. 3-36. The transformer has a separate secondary winding for each SCR and must be designed to withstand the maximum voltage of the load circuit with an adequate safety margin.

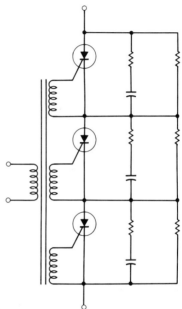

Fig. 3-36. Series-connected SCRs used in triggering scheme for high-voltage inverters.

The circuit shown in Fig. 3-37 is a zero-point switch. It uses discrete components to accomplish a triggering function similar to that of the IC arrangement in Fig. 3-27. As is often the case, there is more here than meets the eye. The circuit is conveniently analyzed by initially assuming that capacitor C2 is *not* present in the circuit.

Suppose that switch S1 is turned on and the circuit is energized during the positive alternation of the ac line voltage. There will be *no* action of any consequence because diode D1 is reverse biased. Commencing with the negative alternation of the ac line, capacitor C1 will charge. The maximum voltage that it can develop will depend on the peak voltage and on the ratio of resistors R1 and R2. This maximum voltage turns out to be about − 40 volts. After the negative alternation has reached its peak, capacitor C1 will *remain* charged at approximately

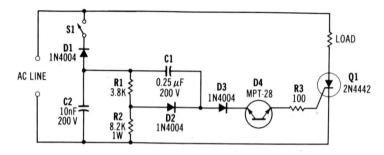

Courtesy Motorola Semiconductor Products, Inc.

Fig. 3-37.   A zero-point switch using discrete components.

-40 volts because diode D2 will then be reverse biased and diac C4 does not have sufficient voltage to fire. However, the voltage appearing across diac D4 *increases* as the negative alternation of the line voltage *declines*. (The line voltage and the voltage stored in capacitor C1 are analogous to two batteries connected in *series-bucking*.) When the line voltage decreases to about 10 volts, the diac will "see" its triggering potential, which is about 30 volts. Note that the SCR gate receives its trigger *before* it is ready to respond—that is, the SCR is triggered while its anode is still *negative* with respect to its cathode. Because of the amount of charge stored in capacitor C2, the SCR gate signal will endure through the zero crossing of the ac line voltage. This action ensures true zero-cross triggering. (Some circuits provide the gate signal slightly *after* zero-crossing and can thereby generate a considerable amount of rfi.)

Let us consider the effect of capacitor C2 in more detail. The circuit action is the same up to the point where capacitor C1 begins to discharge through the diac and the gate circuit of the SCR. But now the instantaneous discharge path will be through D3, D4, R3, the gate of SCR Q1, and capacitor C2. Capacitor C2 becomes quickly charged, thereby reducing the diac voltage and causing it to turn off. Capacitor C2 will then discharge through resistors R1 and R2 until the net voltage across the diac again becomes sufficient to turn it on. This again replenishes the charge in capacitor C2, which again reduces the net voltage across the diac so that it reverts to its blocking state. The time constants are such that *during* an interval embraced by a short time *before* zero-crossing and a short time *after* zero-crossing, a *series* of gate pulses is delivered to the SCR. Thus, the addition of capacitor C2 provides the zero-crossing function with less gate dissipation in the SCR than prevails with a single, prolonged, trigger pulse.

TRIGGERING OF TRIACS

The triggering of triacs is essentially similar to that of SCRs except that *two* trigger pulses per cycle are required. The situation is similar to that of full-wave SCR circuits. However, the circuit implementation of triacs is simpler than that of full-wave SCR arrangements because there is only *one* gate.

In referring to the electrodes of the triac, it would not be relevant to call them an *anode* and a *cathode*. Rather the connections having to do with the load current are known as *main terminal 1* and *main terminal 2*. All voltages at the gate and main terminal 2 are referred to main terminal 1.

The triac will trigger with *either* a positive or a negative gate signal regardless of whether main terminal 2 is positive or negative. It might appear that this would complicate the circuit implementations because of inadvertent triggerings. It turns out, however, that most triacs have greatest gate sensitivity when both the gate and main terminal 2 are positive or when both the gate and main terminal 2 are negative. Other polarity combinations for the gate and main terminal 2 are less likely to result in triggering. On the other hand triacs are available with various gate-polarity sensitivities. It is a matter of selecting the triac that is appropriate for the particular control application.

A simple switching circuit for a triac is shown in Fig. 3-38A. The triac, in response to a low-level gate signal, provides considerable power to the load. In essence this simple arrangement is a solid-state relay. The main advantages over its electromechanical counterpart is that there is no arcing, and there is virtually no maintenance required. Triacs are usually less costly than quality relay-type devices for handling high currents and they generate less rfi and emi. In this circuit it can be seen that the gate is positive when main terminal 2 is positive. Also both electrodes are negative at the same time and there is no opportunity for other polarity combinations. Under these conditions the triac remains turned on through the entire 360 degrees of the power-line cycle.

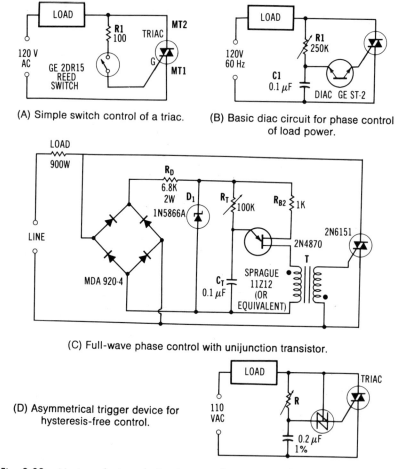

(A) Simple switch control of a triac.

(B) Basic diac circuit for phase control of load power.

(C) Full-wave phase control with unijunction transistor.

(D) Asymmetrical trigger device for hysteresis-free control.

Fig. 3-38. Various devices and techniques for triggering triacs.

Initially, however, the triac may be triggered at *any* point in either the positive or the negative alternation. Therefore, *some* rfi and emi are likely to be produced when the read switch is first closed.

Fig. 3-38B is representative of the basic diac-triac phase-control system. Full-wave control of load current is the prominent feature of such an arrangement. Functionally it is equivalent to a phase-control circuit using a pair of back-to-back SCRs. The fact that the single gate of the triac suffices for control of both power-line alternations obviously brings about simplification. The bilateral triggering characteristics of the diac conveniently complies with the gate characteristics of the triac.

Another commonly used trigger circuit for triacs is shown in Fig. 3-38C. Here a unjunction relaxation oscillator is used in much the same manner that it is often encountered in SCR triggering systems. Of course, *two* trigger pulses per cycle are utilized to impart full-wave control of the triac. The triggering mode experienced by the triac is *not* the same as in the diac circuit of Fig. 3-38B. When main terminal 2 of the triac is positive, *this* circuit delivers a negative signal to the gate, as contrasted to the positive gate signal produced by the diac circuit. Both circuits trigger the gate negatively when main terminal 2 of the triac is negative. For best results in either circuit it is generally wise to *specify* the triggering mode for the triac selected.

The phase-control scheme depicted in Fig. 3-38D is schematically similar to the diac control circuit of Fig. 3-38B. The triggering device is a General Electric ST4 asymmetrical trigger IC. Although *bilateral* like the diac, it has nonsymmetrical breakdown characteristics. The breakdown characteristics of a diac and the ST4 are shown in Fig. 3-39. It might seem that any deviation from the symmetrical-breakdown-voltage feature of the diac could lead to undesirable performance. This is not the case, however, because the symmetrical characteristic of the diac control circuit is objectionable in certain applications, such as in light dimming. When the diac circuit is used, the lamp load cannot be

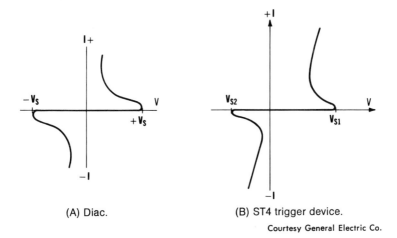

(A) Diac.   (B) ST4 trigger device.

Courtesy General Electric Co.

Fig. 3-39. Breakdown characteristics of a diac and the ST4 trigger device.

gradually turned on. Rather the lamp will suddenly glow at moderate intensity as one attempts to increase light intensity from zero. This is because the timing capacitor is initially uncharged. As the resistance value of control R1 in Fig. 3-38B is slowly decreased, it would be natural to suppose that small triac conduction angles could be gradually brought into play, and that a load could be smoothly adjusted from zero light intensity to brighter levels. However, such is not the case. Let us see why.

Referring to the waveform diagram of Fig. 3-40, which pertains to the diac circuit in Fig. 3-38B, assume that the resistance of phase control R1 is slowly being decreased from its initial maximum-resistance setting. The capacitor voltage will increase until the diac (and therefore the triac) triggers at point A_1. The corresponding triac conduction angle is θ_1. Everything would be fine if the next triggering point were A_2, because the triac would then have a steady-state conduction angle of θ_1 and the lamp load would receive low current. One could then gradually adjust R1 for progressively more lamp current. Unfortunately, however, the triggering advent of A_1 depletes the charge in the capacitor, thereby *lowering* its stored voltage. The capacitor voltage will not, therefore, follow the ideal path shown by the dashed line. So the diac will next trigger at point B rather than at point A_2, which is the symmetrical counterpart of point A_1.

Although point B corresponds to the same diac-triggering voltage as does point A_2, point B occurs at an *earlier* time than point A_2. This results in an abruptly large triac conductance angle θ_2. Thereafter, all subsequent triggerings take place so that conductance angle θ_2 is maintained. That is why the lamp suddenly glows at a moderate brightness

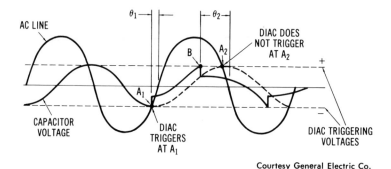

Courtesy General Electric Co.

Fig. 3-40. Waveforms showing diac hysteresis effect.

as resistance R1 is decreased. It is true that the resistance of R1 can then be adjusted for low light intensities. Obviously such an adjustment procedure is a nuisance. This hysteresis is appropriately known as the *snap-back effect* in light dimmers. Paradoxically the use of the symmetrical triggering device, the diac, does *not* lead to symmetrical operation during start-up.

The nuisance value of hysteresis can be even more irritating. Suppose that start-up has been accomplished and the adjustment control is then turned back to produce a low light intensity. If there is a momentary decrease in line voltage, as might occur during turn-on of a heavy appliance, the light could be extinguished and remain so. To turn it on again, one would again have to advance and then to retard the adjustment control.

With the ST4 asymmetrical trigger device, triggering would not occur at point B. Instead a *higher* voltage would have to be reached, but this consumes more time, thereby delaying the firing time of the triac. When the triac does finally fire, the conduction angle, θ_2, is much smaller than indicated in Fig. 3-40. In fact the design of the triggering voltages of the trigger device is such that conduction angles θ_1 and θ_2 are nearly the same. Thus, hysteresis is virtually eliminated.

Hysteresis can also be greatly reduced by the use of a *double-time-constant triggering network*. Such a circuit is shown in Fig. 3-41. Capacitor C2 behaves as a charge reservoir and quickly replenishes much of the charge lost by capacitor C1 when triggering occurs. Fig. 3-42 illustrates that the ideal and actual voltage waves of capacitor C1 are nearly coincident. Thus, the snap-back effect is minimal if not negligible.

The circuit shown in Fig. 3-43 uses a zero-voltage switch, the RCA CA3059, to gate the triac. This versatile IC provides zero-voltage-crossing triggers as indicated in Fig. 3-44. Many different arrangements can be implemented with this device so that burst modulation, feedback control, and various logic and inhibit functions can be realized. Not only

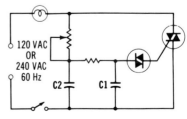

Fig. 3-41. Triac-diac lamp-control circuit with double-time-constant triggering network.

is there no rfi or emi generated by this switching technique, but the triac is subjected to less electrical stress than generally prevails in phase-control triggering.

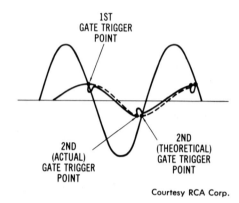

Courtesy RCA Corp.

Fig. 3-42.   Waveform of triggering actions in double-time-constant trigger circuit.

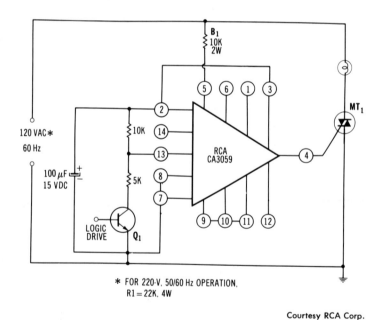

Courtesy RCA Corp.

Fig. 3-43.   Triac triggering circuit using the RCA CA3059 zero-voltage switch.

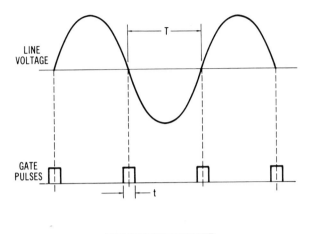

FREQ. (Hz)	T (mS)	t (μS)
60	8.3	100
400	1.25	12

Courtesy RCA Corp.

Fig. 3-44. Waveform diagram showing trigger pulses of RCA CA3059 zero-voltage switch.

FACTORS OF THYRISTORS

A number of factors merit consideration in the interest of successful performance from thyristors. First, it is always wise to try to provide adequate heat removal. In addition to avoiding excessive junction temperatures, another reason for keeping operating temperature low is that all parameters become degraded with rising temperature. It is not an easy task to maintain proper operating conditions if temperature swings are wide. For example, important parameters of triggering voltage and current have sloppy tolerances at best but are worsened with temperature change. Other matters being equal, it is easier to avoid such malperformances as unreliable triggering, false triggering, and commutation problems if the case temperature of the thyristor is kept relatively constant. Of course, when economy is important, the main consideration then is to keep junction temperatures within maximum rating. However, be prepared to cope with challenging circuit demands from your hot thyristors.

Surge Current

A common mistake is to select a thyristor with a current rating apparently properly matched to the steady-state requirements of the load. The thinking is right but does not go far enough. For example, the *cold* resistance of lamp loads may easily be 10 or 15 times less than the hot resistance. See Table 3-4. For a brief time the thyristor may therefore be called on to endure gross overload. When evaluating current capability, it is important to base selection on surge-current ratings as well as the steady-state specifications. A typical surge-current rating curve is shown in Fig. 3-45. Motors are also notorious for their initially high current demands and for their inrush period, which may last much longer than that of lamps. It is both an art and a science to make the most cost-effective provisions for surge-current requirements. However, most specification sheets provide relevant data for this important consideration. Keep in mind that a thyristor with a hot, but safe, junction temperature under *steady-state* operation will be more susceptible to destruction from a turn-on surge than will be a device provided with more effective heat removal, which lowers steady-state junction temperatures.

Inductive Loads

A peculiar commutation difficulty for triacs arises from the effect of inductive loads. Consider the triac circuit shown in Fig. 3-46. Suppose load-current control is attempted through time-delayed gate pulses. This is the technique commonly employed in phase-control thyristor circuits used with essentially resistive loads such as heaters and lamps. With an *inductive* load, however, the load current will lag the line voltage as illustrated by the waveform diagram in Fig. 3-46. As a consequence, when the load current approaches zero and the triac turns off, it will suddenly be impressed with a line voltage that is not zero. Even though the value of this voltage will obviously be far below that corresponding to the maximum blocking voltage of the triac, its *rate of application* (dv/dt) will be exceedingly great. Accordingly, almost instantaneously turn-on will occur and the commutating ability of the triac will be lost.

The usual remedy is the addition of the RC *snubber network* (shown by dashed lines), which limits dv/dt to about 1 volt per microsecond when C_1 is 0.1 microfarad and R_1 is 100 ohms in 60-hertz circuits. How-

Table 3-4. Surge-Current Characteristics of Standard Lamps

Wattage	Rated Voltage	Type	Amps. Steady State Rated Voltage	Hot/Cold Resistance Ratio	Theoretical Peak Inrush (170 V pk) (Amps)	Rated (Lumens/Watt)	Heating Time to 90% Lumens (sec)	Life Rated Hours Average
6	120	Vacuum	0.050	12.4	0.88	7.4	.04	1500
25	120	Vacuum	0.21	13.5	4.05	10.6	.10	1000
60	120	Gas Filled	0.50	13.9	9.70	14.0	.10	1000
100	120	Gas Filled	0.83	14.3	17.3	17.5	.13	750
100 (proj)	120	Gas Filled	0.87	15.5	19.4	19.5	.16	50
200	120	Gas Filled	1.67	16.0	40.5	18.4	.22	750
300	120	Gas Filled	2.50	15.8	55.0	19.2	.27	1000
500	120	Gas Filled	4.17	16.4	97.0	21.0	.38	1000
1000	120	Gas Filled	8.3	16.9	198.0	23.8	.67	1000
1000 (proj)	115	Gas Filled	8.7	18.0	221.0	28.0	.85	50

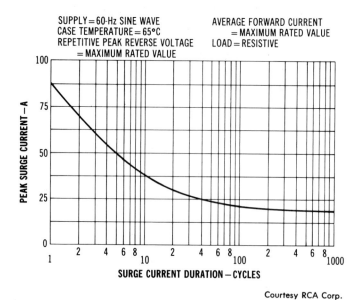

Fig. 3-45. Typical surge-current rating curve for power transistors.

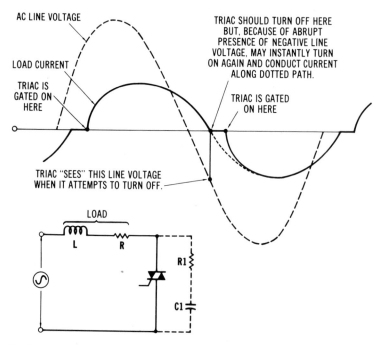

Fig. 3-46. Waveform diagram for triac circuit showing how commutation can be lost with inductive load.

ever, a triac should be chosen that has a *dv/dt* capability of at least 2 volts per microsecond. The purpose of the resistor is to damp any tendency toward ringing of the LC circuit. This would aggravate rather than alleviate commutation problems.

Another commutation problem can be experienced with inductive loads. Because the inductance initially slows the rise rate of the load current, the thyristor may not have attained holding current by the time the gate trigger pulse is over. In this case there will be no sustained turn-on of load current and for practical purposes it will be as if the thyristor has not been triggered. This situation can be alleviated by prolonged trigger pulses or by a series of narrow gate pulses occurring at a high repetitive rate. Sometimes one can experiment with the resistance and capacitance values of the snubber network in order to bring about a situation where the *discharge* of the capacitor through the triggered thyristor will provide sufficient holding until the rising load current eventually takes over.

The application of the snubber network to full-wave SCR circuits is similar to that described for triacs. However, such back-to-back SCRs generally have better *dv/dt* capability to begin with. Even half-wave SCR circuits often have snubber networks. Here the protection may not be needed so much to prevent the effect of inductive lag but to slow transients from the ac line, motors, and other sources.

Blocking and Reverse Voltage

A thyristor must have voltage ratings compatible with the circuit in which it operates. Otherwise it will not be able to *block* the applied voltage in the absence of gate trigger pulses. With an SCR one must make certain that the voltage is never high enough to cause reverse avalanche breakdown. Fortunately this requirement is usually satisfied if the forward blocking state is safely maintained.

Rate of Voltage Application

Compliance with voltage conditions just stated is only a beginning— one must also have some notion of the *rate of applied voltage change* (*dv/dt*) across the main terminals of the thyristor. If this rate is too fast, the thyristor will turn on without a gate signal. This is because the internal junction capacitances will pass a sufficient current to make the thyristor "think" that it has received a gate signal. The mechanism is a simple manifestation of the fact that any capacitance passes current in an amount directly proportional to the rate of change of the impressed voltage. The *dv/dt* ratings are important under two conditions. First, *applied dv/dt* pertains to the simple application of main-terminal voltage to the thyristor. Any thyristor will ultimately self-trigger when such voltage is initially applied too abruptly. *Reapplied dv/dt* is a similar concept but relates to the situation where the thyristor has just been turned off. The significant point is that a certain time must lapse for the thyristor to regain its blocking ability. Such reapplied *dv/dt*, also known as *commutating dv/dt*, varies with the frequency and the magnitude of the load-current pulses.

It is important to keep in mind that the *dv/dt* capabilities of a thyristor can be exceeded by voltages appreciably *smaller* in magnitude than the forward blocking voltage. Indeed, if the rate of change for the voltage is sufficiently great, a voltage with relatively low magnitude can cause false triggering or prevent proper commutation. Such disturbances are often superimposed on the ac power line in the form of transients that, ironically may originate in *other* thyristor equipment.

Dealing with Transients

Remedial measures generally consist of one or more of three basic techniques. First, we should select a thyristor with good *dv/dt* ratings. Second, sharp pulses can be slowed by RC networks known as *snubbers*. And third, various voltage-clipping or energy-absorbing devices can be used. The most applicable of these devices are zener diodes and the General Electric MOV varistor. This device is fabricated from poly-crystalline zinc oxide. Its electrical behavior is similar to that of a back-to-back zener diode arrangement—that is, it exhibits bilateral "crowbar" characteristics similar to the avalanche breakdowns of zener diodes. Its prominent features are its high energy-absorption capability and its speedy response.

Rate of Load-Current Application

We have just discussed thyristor malperformance as a result of slowing the rate of load-current rise following triggering. It is also true that trouble is encountered if this rate, symbolically known as *di/dt*, is too high. The reason for this is that the turn-on process for a thyristor is not instantaneous. Rather some time is required for the current to spread over the cross-sectional area of the semiconductor material. In the thyristor the demand for high load current before turn-on has propagated throughout the junction area in the vicinity of the gate can be destructive, for the current *density* then exceeds safe values. This can result in device deterioration or in quick catastrophic destruction. Although the internal mechanisms and circuit conditions are not quite the same as in secondary breakdown of a transistor, the destructive phenomena in the two devices has a common characteristic. In both instances, localized hot spots are produced that exceed average junction temperature.

A *di/dt* that is too large will not be experienced in simple inductive circuits. Indeed stray inductance in connecting leads often suffices as protection. However, high-frequency circuits working into essentially resistive loads, particularly at high power levels, often require *di/dt* consideration. Another source of excessive *di/dt* can be the snubber network. It has already been pointed out that the discharge of the capacitor through the thyristor can sometimes be advantageously used to *speed di/dt* in inductive circuits where triggering difficulties are experienced. However, the capacitor discharge rate must not be too great, which is another reason for the series resistor. Commutation circuits also have to be monitored closely to be sure that they do not produce excessive current demand.

In addition to taking appropriate circuit precautions, it is necessary to *select* thyristors with *di/dt* ratings suitable for the particular application. As with power transistors, different device fabrications are available for optimization of operating parameters. The law of trade-offs applies in analogous fashion. An example of this is found in the rule-of-thumb relationship for 60-hertz workhorse SCRs:

$$I_{T(AVG)} = \frac{100\sqrt{t_q}}{V_{RM}}$$

where

$I_{T(AVG)}$ is the rated average current in amperes
t_q is the rated turn-off time in microseconds
V_{RM} is the rated blocking voltage in kilovolts

A NEW ERA IN SOLID-STATE POWER

From our brief excursion into heat theory in Chapter 2, it should be evident that the overwhelmingly important factor in establishing power ratings for solid-state devices is the efficiency with which heat can be transferred from a device. This inspires an unending quest for lower overall thermal resistance between the pn junction, or other regions where heat is developed, and the ambient environment. Unfortunately progress becomes slow once we have resorted to massive copper, aluminum, or beryllium-oxide structural elements, good craftsmanship, and, of course, large heat sinks. Semiconductor makers have been obsessed with the possibility that some material might be discovered that would conduct heat tens or even hundreds of times more effectively than ordinary substances. In essence they have found it. The two solid-state devices shown in Fig. 3-47 both have 400-ampere ratings. The disk-like diode in the foreground and the required heat sink behind it represent what has been considered a highly efficient heat-removal technique for convection cooling. The tiny device at the right is also a 400-ampere diode, but it requires no additional thermal hardware. This dramatic comparison represents a technological breakthrough that has been brought about by RCA with its new line of *transcalent* devices, which included diodes, transistors, and SCRs.

Courtesy RCA Electro Optics & Devices

Fig. 3-47. Conventional 400-ampere diode and required heat sink (*left*) compared with new 400-ampere diode developed by RCA.

These new solid-state power devices make use of a so-called *heat pipe* to transfer the heat from the device to the heat sink. The inordinately small heat sink will be shown to be the consequence of the heat pipe. Heat pipes make use of a working fluid (often water) to transfer heat from one region to another. Its use is based on the fact that fluids absorb a tremendous quantity of heat when they evaporate and then, when they condense again, they liberate this heat to the environment. This suggests the nature of the heat pipe. It is a closed system in which the working fluid is converted to its vapor state by the heat from a heat source such as a power transistor. The vapor then expands to a cooler region and liquefies again, liberating its heat in the process. Finally, it is pumped back to the heat source, whereupon the cycle is repeated.

The cutaway view of an experimental heat pipe shown in Fig. 3-48 provides useful insights into the practical aspects of this device.

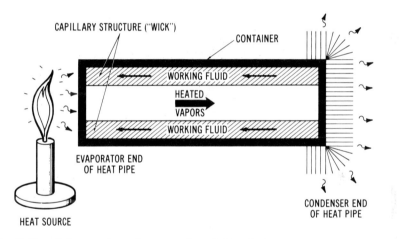

Fig. 3-48. Cutaway view of an experimental heat pipe.

The salient feature of the heat pipe is that no mechanical pump is needed to keep the working cycle going. Rather the "pumping" of the condensed fluid back to the heat source is accomplished by a *capillary structure*, or wick. Such a structure can be woven cloth, fiberglass, wire screen, porous metal or ceramic, or narrow grooves cut longitudinally into the inner surface of the pipe. An example of capillary action is the continuous rise of molten wax in a candle wick. The block diagram of Fig. 3-49 illustrates the sequence of events within the heat pipe.

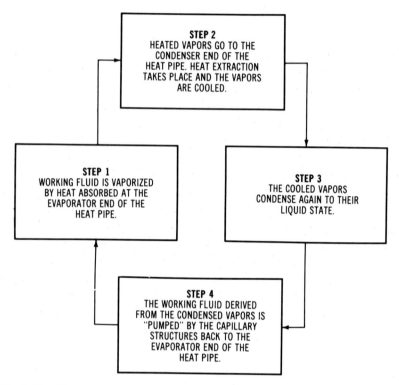

Fig. 3-49. The operational sequence in the heat pipe.

In the RCA transcalent power devices, heat pipes are bonded to both sides of the silicon wafer. Fig. 3-50 depicts such an arrangement for an SCR. Essentially the same fabrication technique is used for diodes and in power transistors. Such structures place power semiconductor ratings in an entirely new range. These transcalent devices are considered exotic and their high cost is justifiable only in military and space applications. The decline in price usually experienced by such new products should soon set in, and these high-performance devices will merit consideration for service in welding, electroplating, motor control, transport vehicles, power distribution, metal refining, induction heating, and other industrial areas where the control of high power levels is involved.

From the foregoing it is clear that anything pertaining to new or improved thermal techniques is bound to be of interest to the user of

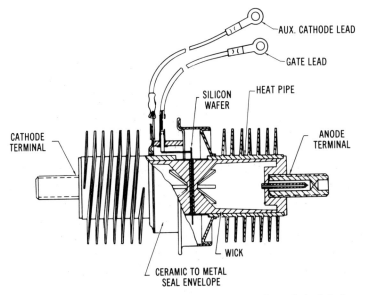

Fig. 3-50. Integrated structure consisting of heat pipes bonded directly to the SCR wafer.

solid-state power devices. Both experimenters and sophisticated de-signers are showing interest in heat-pipe structures that can be used for cooling a large variety of solid-state power devices. Such hardware often takes the form of an elongated box with device-mounting provisions at one end. A fin attaches to the opposite end and may release heat into either still or moving air. The overall system works according to the described principle for the transcalent SCR but has the practical feature that one can assemble the several components in erector-set fashion. The basic idea underlying the use of such a system is that more effective heat removal is often possible than with conventional heat-sink or blower arrangements.

Contemplation of the foregoing discussions shows that the heat pipe actually functions as a *heat pump;* that is, heat is transported from one locality to another. The nature of this transport is that much greater heat exchange occurs than would naturally flow in a heat-conducting substance. Another heat-pumping device—one coming into prominence for specialized applications—is the thermoelectric cooler. This device is actually a semiconductor-diode fabricated to function as a thermocouple and to maximize the Peltier phenomenon. It happens that cause and effect occurring in any thermocouple are *reversible*. Thus, by forcing current in the right direction through such a thermocouple, heat is removed from one side of the junction and is delivered to the other side. If appropriate steps are taken to release the heat to the ambient environment and to prevent it from again impinging on the cold side, an attached heat-generating device, such as a solid-state power device, can be effectively cooled. A practical way of accomplishing this is simply to enclose a portion of the thermoelectric module in thermal insulating material, such as polyurethyne. A principal developer and marketer of thermoelectric coolers is the Cambridge Thermionic Corp.

4

Circuit Applications Using Power Transistors and Power ICs

HEADLIGHT BEAM CONTROLLER

The interesting application of solid-state circuitry shown in Fig. 4-1A automatically lowers the high headlight beams of an automobile in response to the headlights of an oncoming car. The idea is not new, being an adaptation of options offered to buyers of Cadillac and other GM automobiles in the 1950s and 1960s. However, those versions of this scheme used vacuum-tube circuitry as illustrated in Fig. 4-1B. The vacuum-tube model made use of a phototube, a 7851 preamplifier tube, and a 12K5 amplifier tube. The solid-state version is a better performer, is more reliable, and surprisingly utilizes fewer components in a simpler circuit.

When the phototransistor receives incipient light from an oncoming car, its conductivity is increased and it consumes increased collector current through the 22-megohm load resistance. (As is conventionally done in many phototransistor circuits, the base is left electrically unconnected—one might consider the *light* as the source of base "bias.") The increase in collector current of the phototransistor is then amplified in the D32S4/D38S4 first amplifier stage. Additional current gain is then provided by the D41K3 relay driving stage, and the actuation of the relay shifts the headlight circuit from the high beam to the low beam. Two factors contribute greatly to the sensitivity of this circuit. First, the

175

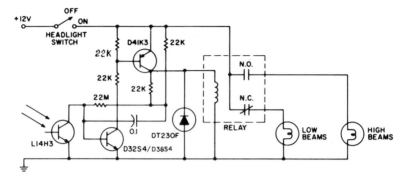

RELAY: 12V, 0.3A COIL; 20A, FORM C, CONTACTS
150A COLD FILAMENT SURGE, RATING.

(A) Solid-state circuit.

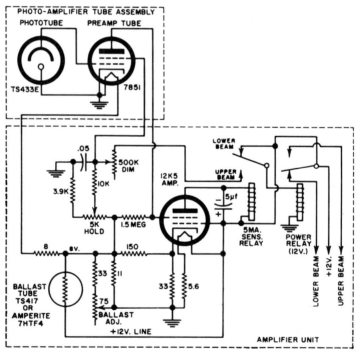

(B) Original vacuum-tube circuit.

Courtesy General Electric Co.

Fig. 4-1.   Automatic beam controller for automobile headlights.

use of a phototransistor rather than a photodiode essentially dispenses with an additional amplifier stage. Second, the D41K3 device is actually a power Darlington with an exceptionally high current gain as shown by the graph in Fig. 4-2. Thus, the power output stage, rather than being a trivial contributor to overall sensitivity, can be said to be a major factor. Also the elimination of discrete amplifier stages in such a direct-coupled system helps attain stability and performance reliability.

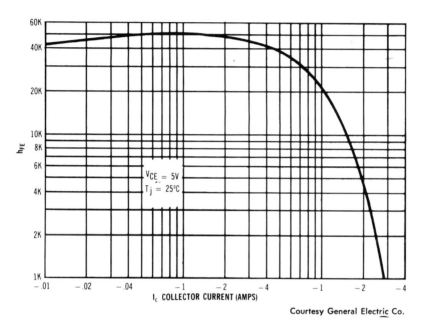

Courtesy General Electric Co.

Fig. 4-2.   Representative curve showing the high current gain of the D41K3.

Of course, the alert experimenter may wish to use all solid-state components by replacing the electromechanical relay with a thyristor or a power transistor. A problem here is that the headlights consume about 16 amperes when the filaments are hot, but cold filaments require a 150-ampere surge. Semiconductor "contacts" will be less perfect than physical contacts insofar as concerns voltage drop. (Balancing this is the practical fact that the resistance of physical contacts increases with age and use. The resistance of metallic contacts increases with temperature—the opposite is true of a semiconductor resistance.)

Inspection of the circuit in Fig. 4-1A will reveal that three cascaded common-emitter stages are involved, counting the phototransistor as the first stage. However, some liberties have been taken with regard to the ordinary configuration of such a circuit. Specifically the positive feedback path provided by the 0.1-μF capacitor endows the circuit with hysteresis. Once activated, the circuit remains in that state for a time; that is, the low beam will remain energized even when the light impinging on the phototransistor disappears. A little thought will indicate that this is necessary in order to prevent objectionable hunting or flashing of the headlights. The DT230F diode connected across the relay coil protects the output transistor from the inductive transients generated when the relay-coil current is abruptly interrupted.

It is recommended that a lens with a minimum diameter of 1 inch be positioned in front of the phototransistor to provide a "viewing" angle of approximately 10 degrees.

A POWER IC INTERCOM

The simple circuit shown in Fig. 4-3 is an intercom system designed around an LM384 IC audio power amplifier. Outputs of up to 5 watts can be obtained under the most favorable conditions, which involve a 26-volt supply, an 8-ohm speaker, and a clip-on heat sink. However, outputs of several watts are readily obtained with a 20-volt power supply and with heat removal provided by several square inches of copper foil on a printed-circuit board. Satisfactory operation at somewhat reduced output is obtained for either 4- or 16-ohm speakers. The quiescent

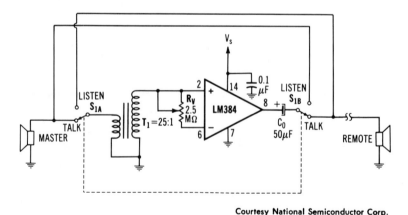

Courtesy National Semiconductor Corp.

Fig. 4-3. Intercom system utilizing an IC audio power amplifier.

current drain from the power supply is low because the output stage operates nearly in Class B. The IC is designed to have a fixed voltage gain of 50, which is why the 25:1 microphone transformer is necessary.

The LM384 has internal thermal limiting and the output is short-circuit proof. It is in a 14-pin dual inline package with six of the pins serving to improve heat transfer to a printed-circuit board. (These six pins are at electrical-ground potential.)

If the master and slave stations are poorly isolated, acoustic feedback may occur with the volume control, R_v, at its maximum setting. If such howling or squealing is experienced, it can often be remedied by reversing either the primary or secondary connections of the microphone transformer. The physical orientation of the speakers is also an influencing factor in such feedback.

ELECTRONIC SIREN

The siren circuit illustrated in Fig. 4-4 is simpler than it appears inasmuch as the transistors and the "op-amp" are integral to a single IC package. The transistor terminals are numbered along with the amplifier symbol. Useful circuits can be conveniently implemented from such an array of active devices in a single IC package. This siren has a peak output capability of about a watt, which will suffice for some signaling requirements. For other requirements, where greater power is needed, this basic circuit can be fed into an audio amplifier with higher power capability. The undulations produced do not simulate the mechanical siren, but the psychological effect is nonetheless similar. The power of a "carrier" frequency is varied at a slow rate. The carrier can be adjusted over a range to produce the greatest psychoacoustical sensitivity. The carrier tone is intensity modulated over an adjustable range of 1 to 7 hertz. Intensity modulation is produced by the switching action of a multivibrator on terminal 3 of the amplifier, which is a tap on the load resistance of the input stage. As part of this load resistance is periodically shorted out by a multivibrator-driven transistor (Q3), the self-oscillating amplifier circuit alternates its gain and power output.

The action of the multivibrator (transistors Q1 and Q2) can be inhibited by placing the single-pole switch in its off position. A sustained, high-level tone will then be produced. The tone format can be controlled by inserting a 2K variable resistance between pins 15 and 3.

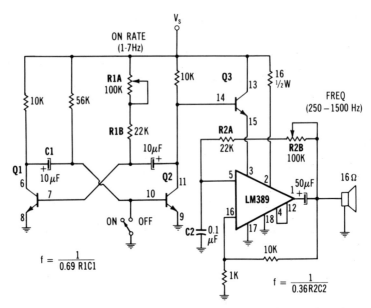

Fig. 4-4.   Solid-state siren circuit utilizing the LM389 IC module.

SIMPLE CIRCUITS USING THE POWER MOSFET

The five circuits shown in Fig. 4-5 are applicable to many more complex systems. Each of these circuits uses the Siliconix VPM1 power MOSFET. The basic idea is that this device can be switched from substantially zero output current to 2 amperes when a positive voltage of 10 volts is applied to the gate. Compared to the base current of bipolar power transistors, the gate current of the power MOSFET can be said to be negligible. The ability to control 2 amperes of current from signal sources providing negligible current is a new dimension of solid-state power manipulation. The freedom from secondary breakdown and thermal runaway, together with high-speed switching capability, lends support to the manufacturer's contention that these devices are destined to displace the bipolar transistor in many applications.

The circuitry for a 40-watt-per-channel stereo amplifier is shown in Fig. 4-6. The schematic diagram in Fig. 4-6A pertains to both amplifier channels. The direct paralleling technique for the two groups of three output MOSFETs is illustrated in Fig. 4-6B. (Sometimes it is expedient

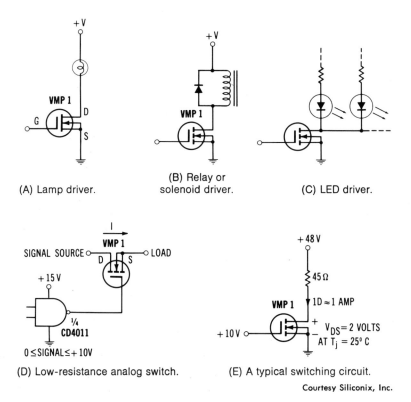

(A) Lamp driver.

(B) Relay or
solenoid driver.

(C) LED driver.

(D) Low-resistance analog switch.

(E) A typical switching circuit.

Courtesy Siliconix, Inc.

Fig. 4-5. Simple circuits using the VPM 1 power MOSFET. The JEDEC designation of this device is 2N6657. The Siliconix VN67AA is an electrically similar device.

to place ferrite beads on the individual gate leads or to connect 1K resistances in series with the individual gate leads to discourage spurious oscillations.) A significant characteristic of the paralleling procedure is that no current-sharing resistances are required. This is because the power MOSFET does not attempt to hog applied current, as do bipolar transistors. Inspection of the output portion of circuit Fig. 4-6A will reveal that these devices are connected in a quasi-complementary-symmetry, push-pull arrangement. The operating mode is Class AB, with just enough idling current to keep crossover distortion at an acceptable level. More recent developments have seen the introduction of many p-channel power MOSFETs. Some of these can be mated with appropriate n-channel devices to form true complementary symmetry circuits (see Chapter 6).

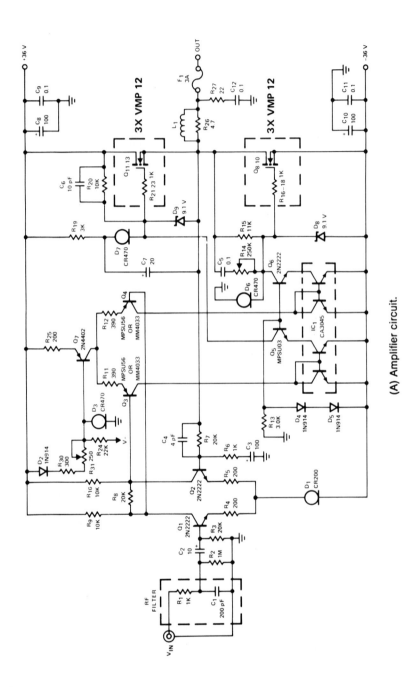

Fig. 4-6. Single 40-watt channel of stereo amplifier with power MOSFET output stage.

(A) Amplifier circuit.

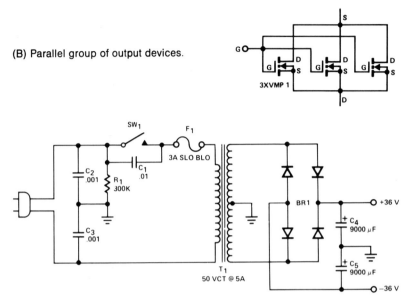

(B) Parallel group of output devices.

(C) Suitable power supply for operation of both output channels.

Courtesy Siliconix, Inc.

Fig. 4-6. *continued*

It is appropriate to mention that many hi-fi enthusiasts feel that the sound from a properly designed power FET amplifier is similar in quality to that previously obtained from well-engineered tube-type equipment and is therefore *superior* to the sound from an amplifier employing bipolar transistors. This is a controversial subject involving much subjectivity and many variables. Quotations of performance parameters usually add more heat than light to any endeavor for objective comparison. However, the power MOSFET design readily lends itself to the use of minimal loop feedback. Additionally this device has an inherent frequency response into the high radio frequencies. Finally, the saturation characteristics are "soft" rather than abrupt as in the bipolar transistor. Whether all this truly leads to enhanced fidelity of sound reproduction is left to the judgment of the reader.

Other features of the amplifier schematic are conventional. The symbols labeled D1, D3, D6, and D7 are constant-current diodes fabricated with junction FETs. They substitute for the commonly used constant-current sources designed around bipolar transistors.

HIGH-EFFICIENCY LAMP DIMMER

The power MOSFET circuit of Fig. 4-7 represents an interesting departure from conventional techniques. Essentially it utilizes a variable-duty-cycle relaxation oscillator to control average drain current in the switched MOSFET stage. Inasmuch as the MOSFET is either on or off, dissipative losses are low. The switching rate is approximately 2 kHz. The duty cycle of the oscillator is determined by the ratio of resistance R1 and R2. Smooth control of light intensity can be obtained from manual adjustment of R2.

On the one hand this control scheme neatly avoids the high dissipation inherent in rheostat circuits, including those in which the "rheostat" is a linearly controlled device such as a bipolar transistor. On the other hand some of the troubles often encountered with phase-controlled thyristors will be either minimal or absent. Specifically one is not likely to experience such shortcomings as hysteresis, lamp flicker, rfi,* tem-

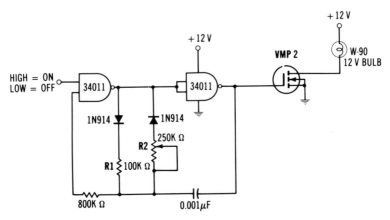

Courtesy Siliconix, Inc.

Fig. 4-7.   High-efficiency lamp dimmer circuit.

---

*A sensitive communications receiver in close proximity to the controller will respond to numerous harmonics of the 2-kHz repetition rate inasmuch as rise and fall times are quite fast. However, the shortcoming of *thyristor*-generated rfi is that the harmonics tend to back up into the ac power line, which then liberally distributes the high-frequency energy via radiation, conduction, and inductive coupling. Such behavior could conceivably plague the MOSFET circuit if it were operated from a power supply connected to the ac line. In general, such trouble will be much less severe than with thyristor circuits because of the "cleaner" operation of the gate oscillator compared to thyristor triggering.

perature effects, or performance variations with different semiconductors. The fact that operation is forthcoming directly from a 12-volt dc supply is also noteworthy. The imaginative experimenter will perceive other potential uses for such a control technique because load control is not limited to lamps. The circuit may also form the basis for other applications, such as audio alarms and switching type regulators. The enable–inhibit function of the free gate terminal enhances the versatility of the circuit. It should also be kept in mind that a number of such power MOSFETs can be directly paralleled for greater load current capability.

POWER IC LAMP FLASHERS

Although lamp-flashing applications are often designed around thyristors, useful circuits can be constructed with other solid-state power devices. The lamp flashers shown in Fig. 4-8 are interesting in that a power IC is utilized to control lamp current. This IC, the LM195/LM295, is drawn schematically as a transistor. It is known as a *synthesized transistor* and has terminals corresponding to the base, emitter, and collector of an npn power transistor. However, this is the extent of the resemblance because the IC device possesses some unique features. One of its salient characteristics is its internal self-protection against excessive current, power dissipation, or temperature rise. For example, inadequate heat-sinking would merely cause an automatic shutdown rather than destruction. The usual SOA curve pertaining to ordinary transistors is not relevant here inasmuch as internal protection guards against secondary breakdown. Moreover, the "base" can be driven up to 40 volts without damage.

Important as the inherent self-protection is, the circuit usefulness of this power IC is mainly due to its tremendous power gain—several microamperes' drive is sufficient for collector current saturation. It is not surprising that these many features are obtained, for this IC contains more than 20 equivalent transistors on a monolithic chip. What may be surprising, though, is its cost competitiveness with conventional power transistors of comparable power and current rating.

The circuit of Fig. 4-8A performs as a multivibrator, with the transistors alternating their conductive states. If both active elements were conventional transistors, one could argue that oscillation would

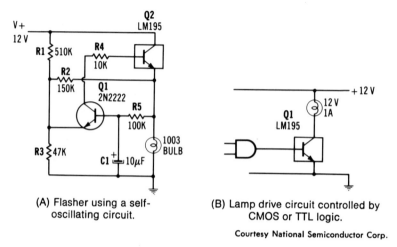

(A) Flasher using a self-oscillating circuit.

(B) Lamp drive circuit controlled by CMOS or TTL logic.

Courtesy National Semiconductor Corp.

Fig. 4-8.   Flasher and lamp drive circuits using a synthesized transistor.

not begin because the application of dc power would find neither transistor forward biased. The synthesized transistor, however, is in some operational respects similar to certain vacuum tubes and FETs. Specifically, when the base circuit of the LM195 is open (transistor Q1 is not turned on), full current is delivered to the lamp, although this is not revealed by the symbol of the LM195. However, there is a constant-current generator connected internally between the collector and base. The constant-current generator behaves as an extremely high resistance and enables the input impedance of the device to be high. At the same time the open-base operating condition is not vulnerable to static charges and is relatively immune to stray pick-up of electrical interference. In any event, when the battery voltage is connected, transistor Q2 immediately supplies lamp current and starts an exponential voltage build-up across capacitor C1. When this voltage becomes sufficient to turn transistor Q1 on, transistor Q2 is turned off and the lamp is extinguished. The current path provided by the base-emitter circuit of Q1 depletes the charge in C1, whereupon transistor Q2 and the lamp turn on again. This cycle is repetitious. As shown, the circuit produces about 45 flashes per minute. (A 500K variable resistance connected across capacitor C1 can be used for adjustment of the duty cycle.)

The lamp circuit of Fig. 4-8B is intended to be controlled by CMOS or TTL logic. Here the high base impedance of the synthesized IC transistor is a marked contrast to the power-demanding input of conventional transistors. Another useful feature of this IC is that several of them can be directly paralleled in order to drive large lamp loads.

A SIMPLE BATTERY CHARGER

Although automotive-type 12-volt batteries may be charged from primitive sources of dc power under duress, certain refinements are generally desired. The battery charger shown in Fig. 4-9 has features that may not be initially apparent from the simplicity of the circuit. This charger is capable of sustaining a three- or four-ampere charge rate until the battery attains a substantially full charge, whereupon the charge rate is automatically reduced to a much lower level. If during the charging process the ac power should be switched off, or otherwise interrupted, there will be virtually no drain imposed on the battery, even though it remains connected to the charger. The charger will not be damaged by either shorts or reverse-polarity battery connections. Although most or all of these features have been accomplished by various other circuits, the circuit of Fig. 4-9 is favorable with regard to cost, ruggedness, and performance.

The basic configuration of this charger is that of a voltage-regulated power supply. Two differences from ordinary regulator operation arise from the fact that a "live" rather than an essentially resistive load is

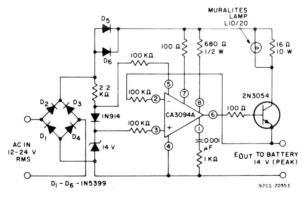

Courtesy RCA Corp.

Fig. 4-9. Battery charger with automatic rate control and other protective features.

involved and that the CA3094A IC is not an ordinary operational amplifier. Specifically terminal 5 provides the function of current-proportional gain control. That is, the output of this IC is governed not only by the voltage difference sensed at its differential input amplifier (terminals 2 and 3) but also by the current flowing into terminal 5. (Internally this current controls the supply of current to the differential amplifier and thereby the transconductance of its pair of transistors.) Yet another feature of this IC is that its current and power capability definitely place it in the category of power devices. It can deliver a steady current of 100 mA and a peak output current of 300 mA. This also corresponds to a sustained load power of 3 watts. This IC is thus able to drive directly a medium-power transistor such as the 2N3054. The internal dissipation of this device is only a few microwatts in the quiescent state.

The salient operating feature of this charger is that it samples the battery voltage and shuts down the charging current when the battery develops 14 volts across its terminals. This is accomplished via the 100K resistance connected to terminal 2, the inverting input. In this regard the op amp behaves as a voltage comparator, with the 14-volt zener diode providing the reference voltage. On the other hand, *during* the charging process while the battery voltage is *below* 14 volts, charging current passes to the battery.

FAST CHARGERS FOR NICKEL-CADMIUM BATTERIES

The advent of the rechargeable nickel-cadmium battery has resulted in cordless freedom for many electrically operated products. Included are shavers, soldering irons, transceivers, hand drills, toothbrushes, food mixers, carving knives, vacuum cleaners, and clothes brushes. Although the convenience gained through the use of battery power is obvious, a major *inconvenience* has also accompanied this new mode of operation. It is commonly required that a charging process of about 15 hours be applied to restore a discharged battery to full capacity. Unfortunately the nickel-cadmium cells are not as amenable to "quick charging" as are the lead-acid cells used in automobile batteries. It has, however, been determined that fast charging is practical providing that the charging current is turned off (or reduced to a "trickle-charge" rate) as soon as a fully charged condition is reached. Otherwise there will be an abrupt rise in cell temperature together with the emission of gas. This may result in rupture of the cell and is otherwise undesirable because it tends to shorten the life span of the cell in terms of the total number of charge-discharge cycles that the cell can provide. Venting is not a practical solution, inasmuch as the lost electrolyte diminishes the capacity of the cell.

Two experimental circuits that enable considerable reduction in charging time are shown here. It appears that a 1-hour charging period is feasible in some cases. Under laboratory conditions even greater acceleration of the charge time is possible. The difficulty involved is the *detection* of some cell parameter that indicates a full charge. Cell *voltage* has thus far proved more reliable than other parameters, such as temperature or impedance. Fig. 4-10 is a plot of cell voltage as a function of charge time for ordinary nickel-cadmium cells and for a special type nickel-cadmium cell provided with a *third* electrode. Exceptionally stable and precise detection is needed for determining a full-charge condition in ordinary cells. Somewhat balancing the detection problem, it has been found that all nickel-cadmium batteries are more receptive to *pulsed* current than to straight direct current. This is fortunate due to the efficiency and controllability of electronic pulsed-energy systems.

The circuit of a 25-kHz battery charger for conventional nickel-cadmium batteries is shown in Fig. 4-11. This system consists of two functional sections, one being the 25-kHz power supply and the other comprising the full-charge detection circuitry. The power supply will be discussed first.

The 25-kHz pulsed current is generated by a blocking oscillator utilizing the MJ3010 power transistor, the three-winding transformer, and associated circuitry. The dc operating power for this blocking oscillator is obtained from the 120-volt, 60-Hz line via the MDA920 bridge rectifier and the 8-μF filter capacitor. (When using this charger, the usual precautions pertaining to nonisolated ac equipment should be observed.)

The 25-kHz blocking oscillator operates by virtue of the regenerative feedback from the primary winding of the blocking transformer to the feedback winding. The MJ3010 power transistor is alternately driven to saturated conduction, then to cutoff. However, its oscillatory period is not governed by an RC time constant, as with many blocking oscillators. In this circuit a second transistor, the MPS6512, senses the rising primary current as a voltage drop across the 3.0-ohm resistance. When this voltage drop becomes 1.2 volts, the MPS6512 control transistor is forward biased and conducts sufficiently to divert the base drive of the oscillator transistor to ground. The current ramp terminates and the energy stored in the magnetic field of the transformer is released as a pulse of current through the fast-recovery diode, MR880, into the battery. (The 1.2-volt sensing level represents the sum of the forward biases of one of the MSD6100 diodes and the base-emitter diode of the MPS6512 control transistor.) The cycle then repeats, with a new current ramp produced in the primary winding of the transformer.

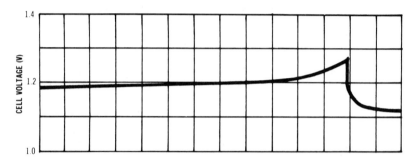

(A) Attainment of fully charged condition in an ordinary nickel-cadmium battery.

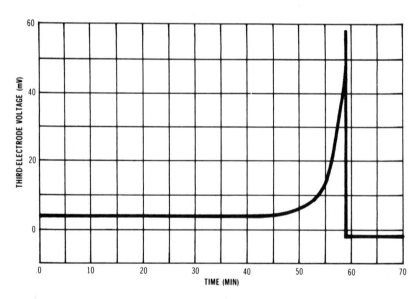

(B) Attainment of fully charged condition in a special third-electrode nickel-cadmium battery.

Courtesy Motorola Semiconductor Products, Inc.

Fig. 4-10. Cell voltage changes for ordinary and special-type nickel-cadmium batteries.

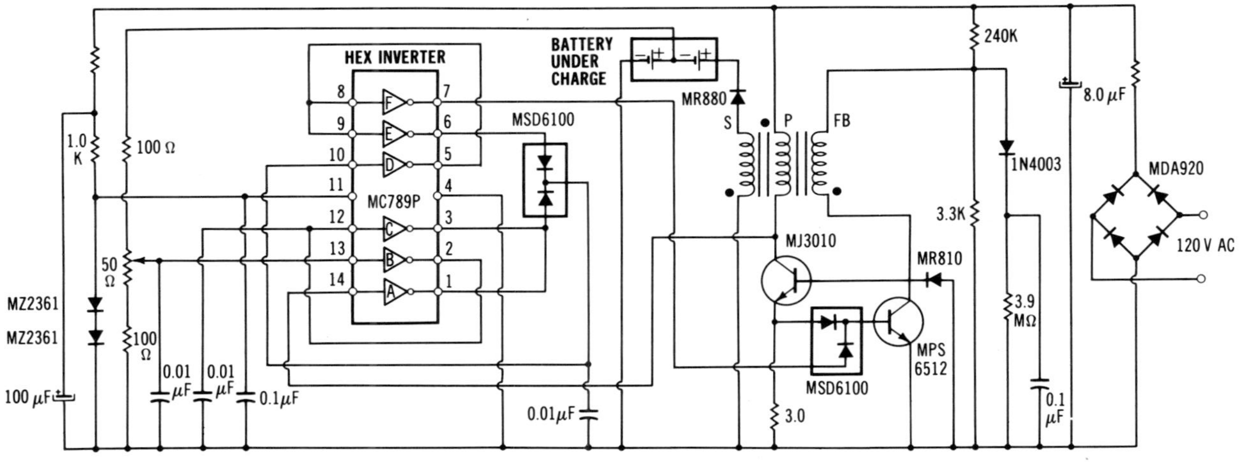

Fig. 4-11. Fast charger for conventional nickel-cadmium batteries.

191

The presence of the MSD6100 diode pair is due to the fact that the control transistor is also used to sense the fully charged condition of the battery and it is desired to keep the two sensing functions isolated from each other. In this second sensing function the control transistor inhibits oscillation as long as the terminal voltage of the sampled battery cell remains appreciably higher than it is when the cell is less than fully charged.

A transient-suppressor network is connected to the base of the oscillator transistor to protect it from destructive spikes during the turn-on portion of the cycle. This network consists of the 1N4003 diode, the 3.9-megohm resistance, and the 0.1-μF capacitor.

The sensing circuit is configured around the MC789P hex inverter. This RTL integrated circuit provides six independent inverter stages admirably suited to the requirements of this charger. The cell voltage is monitored through the 50-ohm potentiometer by inverter B. The base-emitter diode of this stage provides the threshold level as well as the temperature compensation for the sensing circuit. Amplification of the output from inverter B is provided by inverter C. Inverters D and E comprise a flip-flop. Through their action the 25-kHz oscillator is kept inoperative after the battery becomes fully charged. Inverter F is a final buffer stage, which "reports" the status of the sensing circuit to the MPS6512 control transistor in the 25-kHz oscillator circuit. Note that we have not yet accounted for inverter A. Inverter A short-circuits the output of inverter C during the time interval in which the battery is accepting charging current. It accomplishes this indirectly by sampling the collector voltage of the MJ3010 oscillator transistor. Because of this circuit action, the sensing of the cell voltage can control the charger only when no current pulses are being delivered to the battery. This provides a more realistic monitoring of the charge status of the battery.

The hex inverter module receives about 2.5 volts of regulated dc operating power from the two series-connected MZ2361 stabistor diodes. Although forward conducting, these diodes regulate in a manner similar to that of zener diodes. The 0.01-μF capacitors connected to the hex inverter provide required electrical noise immunity to the sensing circuit.

When the charger is placed in operation, some initial experimentation will be necessary to determine the proper setting of the sensing adjustment potentiometer. After the correct setting has been found for a given type of battery, good performance with fast charge rates and reliable turn-off despite changes in line voltage and temperature should prevail. If, however, the special nickel-cadmium batteries with third-electrode sensing provision are available, the charger to be described next is recommended.

The fast charging circuit shown in Fig. 4-12 is intended for nickel-cadmium batteries with third electrodes. As shown in Fig. 4-10, the third electrode provides a much larger voltage *differential* for the electronic sensing of the fully charged condition. Accordingly the charger of Fig. 4-12 employs a 25-kHz current pulsing circuit similar to that of Fig. 4-11, but it is able to provide the turn-off function with a simpler sensing circuit. At the same time the simplicity of the sensing circuit incorporating the MD8001 monolithic transistor pair can be deceptive. Attempts to simulate this circuit with discrete transistors are bound to prove unsuccessful. This is because only monolithic fabrication can provide two requirements of this circuit: *closely matched transconductance* and *tight thermal coupling* between the transistors. When these requisites are met, the thermal coefficient of the sensing circuit can be calculated as a

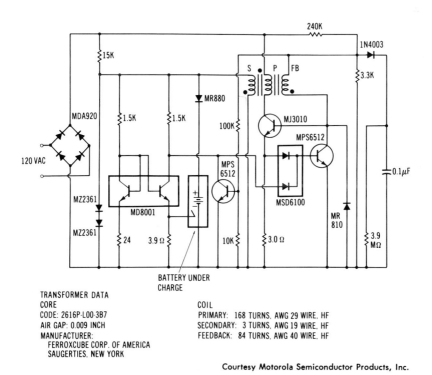

TRANSFORMER DATA
CORE
CODE: 2616P-L00-3B7
AIR GAP: 0.009 INCH
MANUFACTURER:
    FERROXCUBE CORP. OF AMERICA
    SAUGERTIES, NEW YORK

COIL
PRIMARY: 168 TURNS, AWG 29 WIRE, HF
SECONDARY: 3 TURNS, AWG 19 WIRE, HF
FEEDBACK: 84 TURNS, AWG 40 WIRE, HF

Courtesy Motorola Semiconductor Products, Inc.

Fig. 4-12.   Fast charger for special nickel-cadmium batteries with third electrode.

function of the emitter and collector resistance. Thus, the thermal coefficient of the sensing circuit can be made to compensate that of the battery. Qualitatively the left-hand transistor may be considered the forward-bias supply for the right-hand transistor that operates as a constant-current source. The collector voltage of the latter transistor therefore behaves as if derived from a programmable zener diode, with the programming being imparted by the voltage sensed at the third electrode of the special nickel-cadmium battery.

As with the previously described charger, it is desirable to sample charge status of the battery *between* current charging pulses only. This is accomplished by the MPS6512 transistor that is turned on whenever the positive current pulses are delivered to the battery, thereby diverting the output of the sensing circuit. In this way the third-electrode voltage becomes more representative of the true charge state of the battery, and the full-charge turn-off performance of the charger is improved in both accuracy and reliability.

The same precautions with regard to the nonisolated power-line connections for the previous circuit also apply for this charger. The MJ3010 power transistor in both chargers should be equipped with a suitable heat sink inasmuch as one may wish to experiment with average charging currents of up to 4.8 amperes for the commonly encountered 1.2-ampere–hour nickel-cadmium cells.

INDUCTIVE-DISCHARGE IGNITION SYSTEM

There are several ways to classify the many different types of ignition circuits that have been used for automobiles. Whether solid-state or "conventional," almost all ignition systems can be neatly described as *inductive-discharge* types or *capacitive*-discharge types. Of the two types the latter is probably the more elaborate and is capable of superior energy delivery to the spark plugs even at high speeds. It is interesting, however, that the automobile manufacturers decided to deploy the inductive-discharge type as standard equipment. It may well be that the picture will change in the next several years, but it does appear that a wise decision was made. The solid-state inductive-discharge type provides a number of superior features with respect to the conventional (Kettering) ignition systems used for so many years. The even better performance of the capacitive-discharge system is more meaningful to the hot-rodder or the perfectionist than to the average motorist. From the viewpoint of the manufacturers the solid-state inductive-discharge type tends to be less costly and may be more reliable for everyday motoring than capacitive-discharge ignition systems. The capacitive-discharge system generally develops higher voltage and is more demanding with regard to quality of insulation. More consideration is usually required with regard to routing of high-voltage cables, selection of the ignition coil, and choice of spark plugs. Conversely the inductive-discharge system more readily substitutes for the conventional ignition system.

The Chrysler Corporation began equipping its cars with an excellent inductive-discharge ignition system in the 1973 models. In addition to reliable solid-state circuitry, the breaker points and "condenser" of conventional systems were replaced by a practically maintenance-free magnetic *reluctor*. The reluctor is a toothed steel wheel that rotates in close proximity to the pole tip of a permanent magnet. Associated with the magnet is a pick-up coil. The flux variations thread through the pick-up coil as a result of a tooth moving past the pole tip of the magnet. This induces an emf that in turn actuates the electronic circuit and produces a precisely timed spark for the appropriate cylinder. There are as many teeth on the reluctor as there are cylinders in the engine. The design of the high-voltage distributor remains "conventional."

It is well to point out here that the inductive-discharge classification does not pertain at all to the reluctor and pick-up coil. Indeed inductive-discharge ignition systems have been implemented with retention of the breaker points of conventional ignition systems. Inductive-discharge ignition systems are so-called because the energy for the production of the high-voltage spark is alternately stored and "drained" from the *inductance of the ignition coil.* That is, the required energy exists in the magnetic field of the ignition coil before its conversion to thermal energy in the gaps of the spark plugs. By this definition it is evident that conventional ignition systems are also inductive.

In the circuit of Fig. 4-13 power transistor Q5 is controlled in such a way that a current path is maintained through the primary of the ignition coil except when a cylinder is to fire. At such time, Q5 is abruptly turned off and the magnetic field previously supported by the primary current collapses. This induces the high voltage in the secondary circuit, which, of course, includes the appropriate spark plug. This timed spark-plug firing occurs as the result of a reluctor tooth approaching and

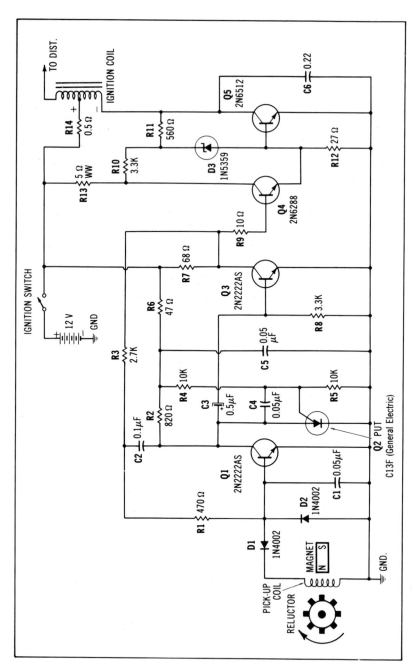

Fig. 4-13. The Chrysler-type electronic ignition system.

receding from an exposed pole of the permanent magnet. Specifically, as the reluctor tooth completes its position of alignment with the pole tip and begins to recede, the induced voltage in the pick-up coil changes from positive to negative polarity, thereby establishing a circuit through diodes D1 and D2. This also turns off normally on transistor Q1, which in turn delivers a positive pulse through capacitor C3 to the base of normally off stage Q3. Finally, both driver Q4 and output stage Q5 are turned off from their normally on states. The resultant interruption of primary current in the ignition coil induces the high voltage for spark-plug firing in the secondary winding.

The function of programmable unijunction transistor (PUT) Q2 was not included in the sequence of events just described. The cascade of switching stages are not all directly coupled. Rather ac coupling is provided between the first stage (Q1) and the second stage (Q2) by capacitor C3. It is not desirable to cascade too many stages through direct coupling because of drift and temperature problems. The ac coupling solves this problem. In a pulse amplifier it is desirable that the charge in coupling capacitors be depleted during the interval between pulses. This is accomplished by the firing of PUT Q2. Note that Q2 is not immediately ready to fire at the instant that transistor Q1 opens and provides it with anode voltage. Rather a *time delay* is imposed because of the charge in capacitor C4. At the elapse of this time delay, Q2 fires and dissipates much of the residual charge in coupling capacitor C3. When Q1 returns to its on state, transistor Q2 is commutated back to its off state. The entire circuit is then ready to generate another burst of high voltage for the next spark plug in the firing format of the engine.

POWER IC ADJUSTABLE CURRENT REGULATOR

Although the stabilization of voltage generally comes to mind when dealing with regulated dc power supplies, there are many applications in which the regulation of current provides more meaningful measurement data, more desirable performance, or better control. Many electromagnetic devices fall into this category, including relays, motors, solenoids, and deflection yokes. For example, the pull-in and drop-out current of relays is conveniently determined with a programmable constant-current source. Certain dc motors can be relied on for predictable and repeatable performance when operated from constant current. Constant voltage is less meaningful inasmuch as the basic motor torque is a *current* function and the motor current is not tightly stabilized by operating from a constant-voltage source because the resistance of the copper windings is temperature dependent. Moreover, the voltage drop across the brushes is not closely predictable and varies erratically with load conditions and with aging.

Electrochemical processes, such as battery charging and electroplating, are similarly current functions. Here, too, better measurements and more accurate control result from the use of current, rather than voltage, regulators.

From consideration of the basic physics of other devices, a number of tailor-made applications are often found for current regulators. For example, the control of thermionic emission in such equipment as electron microscopes and mass spectrometers can lead to improved performance as well as extended filament life. Again, current, rather than voltage, is the more relevant parameter in heating such filaments. Finally, some semiconductor devices such as zener diodes and some microwave diodes can best be made to perform their functions when used with a constant-current source of power.

The current regulator shown in Fig. 4-14 can be adjusted to provide 0 to 3 amperes of load current. It consists basically of two adjustable three-terminal voltage regulators with versatile circuit and operational features. The LM150 functions as the current regulating IC in this circuit arrangement, which can provide up to 3 amperes of regulated output current. The LM117 operates as an external reference, which in *conjunction* with the internal voltage reference of the LM150 enables adjustment of load current down to zero. An auxiliary negative supply is needed, but it has to provide only 10 milliamperes of current. The input voltage can be between 10 and 35 volts.

This deceptively simple schematic, consisting of two "blocks" and several resistances, should not be construed to be anything less than a high-performance system because these blocks are the equivalent of a total of more than 60 discrete transistors. Included are such features as thermal regulation, thermal overload protection, automatic current limiting (at greater than the guaranteed 3-ampere output current), and SOA protection of the output transistors.

Be sure to use the indicated ICs—most other three-terminal IC regulators are optimally applied where fixed output voltage or current

Fig. 4-14. An 0–3-ampere adjustable current-regulated power supply.

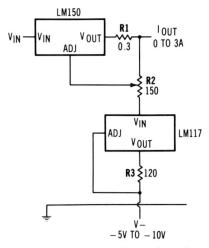

Courtesy National Semiconductor Corp.

is needed or where a limited adjustment range suffices. The LM150 and LM117 ICs require much less ADJ terminal current than do "fixed" types. This makes it easier to adjust down to zero current. It also enhances regulation and the linearity of the resistance-to-current relationship, particularly at low load currents.

A 1-AMPERE, ± 15-VOLT TRACKING REGULATOR

A valuable addition to the designer's or experimenter's laboratory is a single source of regulated ±15 volts, because many ICs require this format of operating power. Regulated power supplies constructed with discrete active devices have served this purpose satisfactorily. However, it is startling to contemplate the simplicity afforded by modern ICs, such as the three-terminal voltage regulator. When these devices are employed, cost and parts count are dramatically reduced. But best of all, the performance exceeds that generally attainable with discrete devices.

The dual-voltage tracking regulator shown in Fig. 4-15 clearly illustrates this new circuit technique. Conspicuous by their absence are several transistors and their associated passive components. There is no voltage reference circuit, and no protective techniques other than the diodes connected across the outputs. These circuit functions are incorporated *within* the ICs, and we need not be concerned about them.

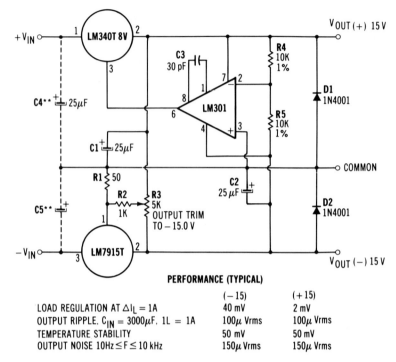

PERFORMANCE (TYPICAL)

	(−15)	(+15)
LOAD REGULATION AT $\Delta I_L = 1A$	40 mV	2 mV
OUTPUT RIPPLE. $C_{IN} = 3000\mu F$, $1L = 1A$	100μ Vrms	100μ Vrms
TEMPERATURE STABILITY	50 mV	50 mV
OUTPUT NOISE $10Hz \le F \le 10\ kHz$	150μ Vrms	150μ Vrms

*RESISTOR TOLERANCE OF R4 AND R5 DETERMINE MATCHING OF (+) AND (−) OUTPUTS

**NECESSARY ONLY IF RAW SUPPLY FILTERS CAPACITORS ARE MORE THAN 3" FROM REGULATORS

Courtesy National Semiconductor Corp.

Fig. 4-15.   A 1-ampere, ± 15-volt tracking regulator.

Thus, the usual headaches associated with making regulated supplies stable and reliable are avoided.

It is true that the unregulated dc sources are not shown in Fig. 4-15. However, these can be quite primitive, just so the voltage and current requirements are met. For example, they can consist of a common transformer with two 15- to 18-volt secondaries, appropriate bridge rectifiers, and about a 3,000-μF filter capacitor for each supply. As can be seen from the performance specifications, the regulation and ripple suppression of this regulator are exceedingly good.

The LM340T and the LM7915T are similar ICs except that the former is a *positive* voltage regulator and the latter is a *negative* voltage regulator. As employed in the circuit of Fig. 4-15, the LM7915T is the *master* supply, and the LM340T is the *slave* supply. The numbering of the three leads is not the same for the two ICs. (For the LM7915T, no. 1 is the ground, or "adjust" terminal; no. 2 and no. 3 are output and input terminals, respectively. For the LM340T, no. 1 is the input terminal; no. 2 is the output terminal; and no. 3 is the ground or "adjust" terminal.) Note also that the 8-volt version of the LM340T is used. The actual output of the LM340T is boosted to 15 volts or, more precisely, to the output voltage of the LM7915T master supply. The LM301 op amp performs the required polarity inversion needed to bring about this voltage boost. The op amp also enhances the dynamic performance of the positive supply.

A POWER MOSFET 200-kHz SWITCHING REGULATOR

For the past decade, 20-kHz switching regulators have been considered ideal. In the vicinity of this frequency, one obtains a pragmatic balance of cost, performance, and packaging compactness. At lower frequencies there is the possibility of disturbing acoustic noise. At higher frequencies a suitable bipolar switching transistor is likely to be expensive. And other devices, such as the filter capacitors and the free-wheeling diode, have posed stumbling blocks with regard to speed and losses. However, for further improvements in the power-to-weight ratio, it has always been desirable to go beyond the present 20- to 25-kHz switching rate. Also, at higher switching rates, certain emi problems tend to be reduced as higher frequencies are more readily suppressed with practical shielding and filtering techniques.

Switching regulators can be economically constructed to operate at higher switching rates than what has hitherto been considered feasible. The advent of power MOSFETs with power capabilities of more than 10 amperes is likely to revolutionize previous notions of design boundaries. Also, Schottky diodes and vastly improved filter capacitors increase the practicability of such high-frequency regulation. The regulator circuit shown in Fig. 4-16 is capable of providing 10 amperes regulated at 5 volts. The 710 voltage comparator has higher speed capability than most op amps and is deployed here as a self-excited oscillator and pulse-width modulator. Oscillation occurs via the positive feedback path provided by resistance R9. The noninverting terminal (+) of the voltage comparator samples both the dc and ac components of the output and

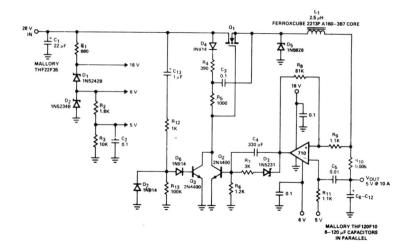

Courtesy Siliconix, Inc.

Fig. 4-16.   A 200-kHz switching regulator.

is driven to produce a voltage nulling action at its output. Such nulling is achieved by utilizing more or less of the roughly triangular "ripple" voltage. The nulling is the consequence of duty-cycle, or pulse-width, modulation. If the dc output voltage tends to drop, *more* triangular voltage is utilized; switching-on time becomes *longer* and the dc output voltage of the regulator increases again to its set level. The converse is true when there is a tendency for output voltage to rise, and *less* of the sampled triangular wave is utilized to drive the voltage comparator to saturation. *Shorter* on-time then causes the dc output voltage to decrease to its set level. Close voltage regulation of the dc output takes place because the error-signal nulling action is referenced to the constant 5-volt source provided by zener diode D2. (No doubt the experimenter can devise a stabler reference source than the one shown. The objective of the overall design was focused on the 200-kHz switching rate and the resulting small physical size of the major components.)

Transistor Q3 and its associated components form a soft start-up circuit. When the regulator is first energized from the 28-volt dc source, capacitor C13 passes sufficient charging current to turn on transistor Q3, which is normally off. This in turn clamps the collector of transistor Q2 to ground, thereby depriving switching element Q1 of enhancement bias. The charging of capacitor C3 gradually alleviates this condition, and the output voltage of the regulator rises smoothly from 1 to its rated output level of 5 volts. In this way the possibility of a momentary, but load-endangering, overshoot is prevented.

Switching element Q1 is intended to be the Siliconix VN84GA 12.5-ampere power MOSFET. However, the unique characteristics of these devices enable convenient paralleling of power MOSFETs with lower individual current ratings. For example, a group of five type 2N6657 units can be mounted on a common heat sink. The performance obtained will be similar to a single VN84GA inasmuch as the 2N6657 has a 2-ampere rating. Unlike bipolar transistors, these devices do not hog current or participate in thermal runaway. It is not necessary to insert ballast resistances, as commonly required when paralleling bipolar type transistors. Nor is it necessary to be concerned over available drive because for practical purposes the power MOSFET exhibits infinite input impedance when deployed in a circuit such as Fig. 4-16. Of course, packaging, cost, and production considerations favor the use of a single device rather than its paralleled equivalent.

SOLID-STATE CIRCUIT BREAKER

Fuses, thermostatic switches, and electromechanical circuit breakers all are used for the protection of electrical and electronic equipment. There are situations, however, when these protective devices are not altogether satisfactory. Their shortcomings may involve sluggish action, indeterminate trip setting, bulky physical dimensions, cost, or the inconvenience of needing replacement.

It has long been known that automatic interruption of current or power could be accomplished with solid-state devices. However, until recently it has not been easy to obtain electrical ruggedness together with respectable voltage and current capabilities. The circuit shown in Fig. 4-17 is superior to earlier versions of electronic circuit interruptors. The improved performance is largely attained through the use of the DTS-4039 triple-diffused Darlington transistor. Also, the parts count is lower than it initially appears because the three op amps are part of a low-cost, quad IC. This circuit breaker safely operates at voltages as high as 300 volts and can interrupt currents up to 8 amperes. (These are *simultaneous* ratings.) The response time is on the order of 300

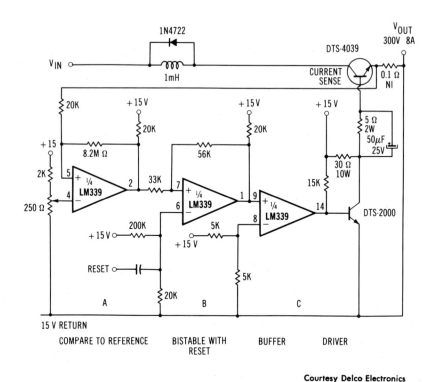

Courtesy Delco Electronics

Fig. 4-17. Schematic diagram for the solid-state circuit breaker.

milliamperes per microsecond. Such rapid interruption can be a vital factor in saving sensitive devices from the damage or destruction often caused by short circuits when other protective devices are used.

Op-amp A is employed as a voltage comparator. It compares the voltage monitored across the 0.1-ohm "current-sense" resistance with a voltage provided by the 250-ohm adjustment potentiometer. The output of this comparator swings from 0 to +15 volts when the sense voltage exceeds the preset voltage from the potentiometer. The 8.2-megohm resistance connected between input and output of this stage produces a small amount of positive feedback. This in turn introduces sufficient hysteresis in the response of the comparator to discourage instability at the trip point.

Op-amp B has the circuit configuration of a binary flip-flop. A positive step from the voltage comparator causes the output of this binary stage to abruptly switch from zero to $+15$ volts. Note the provision for a *reset* pulse. A narrow positive pulse with an amplitude of several volts suffices to restore operation after the fault has been cleared. The reset switch can be remotely located for convenience.

Op-amp C is used as a buffer or driver amplifier for the DTS-2000 drive transistor. The DTS-2000 is actually a Darlington type and is employed to divert to ground the 400 mA or so of current required by the base of the "interruptor," the DTS-4039. Thus, when the output of op-amp C swings positive, the DTS-4039 is deprived of forward bias and therefore opens the load circuit.

In order to simplify the schematic layout of the circuit diagram, the $+15$-volt supply for the op amps and for bias is not shown. It can be an IC module with about 450-mA current capability.

The 50-μF capacitor in the base-bias circuit of the DTS-4039 transistor makes the turn-off process more abrupt. Its function is analogous to the speed capacitors used with digital logic gates. The 1-mH inductor prevents an instantaneous surge of high current through the DTS-4039 transistor in the event of a short circuit at the load. The IN4722 diode across this inductor protects the DTS-4039 from the voltage transient, which would otherwise be generated during turn-off.

In normal operation the DTS-4039 transistor is in saturation. This results in relatively low power dissipation (about 12 watts, maximum) and makes the heat-sink requirement minimal. The Delco 7277151 heat sink is adequate for ambient temperatures up to at least 75°C. Because of the minimal thermal dissipation requirements and the single IC, the packaging of this circuit can be made reasonably compact.

A 115-VOLT, 60-HERTZ INVERTER
FOR AN AUTOMOBILE

Inverters are useful as sources of ac power when the primary source is a battery, as is the case for motor vehicles, boats, and aircraft. Another widely used application for the inverter is as the basic energy converter in regulated power supplies. When followed by a rectifier, the inverter becomes a dc transformer that allows voltage and current levels available from the primary dc source to be changed to more convenient levels. There are many high-performance inverters available operating at high frequencies, achieving high efficiency, and making use of sophisticated devices. However, there remains a continuing demand for simple inverters that can be conveniently constructed from low-cost semiconductors and other components.

The simple inverter shown in Fig. 4-18 provides practical functions and is inordinately easy to construct. It delivers about 50 watts of power, which is ample for the operation of electric shavers, small soldering irons, recorders, many radios, appropriately rated fluorescent desk lamps, and small portable phonographs. The heart of the unit is a 50-watt, 24-volt, center-tapped filament transformer, which should be readily attainable.

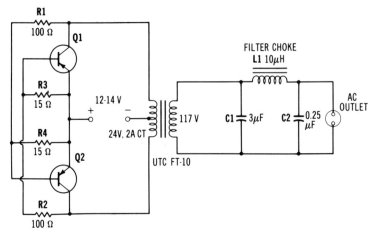

Courtesy Motorola Semiconductor Products, Inc.

Fig. 4-18. A 115-volt, 60-hertz inverter for operation from an automobile battery.

The original germanium transistors specified for this circuit were type 2N176. However, many workhorse types will perform well in this circuit. Suggested substitutes are the Motorola types 2N1535, 2N1540, 2N1545, 2N5894, or 2N3614. Some experimentation with the value of feedback resistors R1 and R2 will probably be required for optimum starting and operating performance. (Silicon pnp transistors such as the Motorola MJ490 will work but *less* efficiently than germanium units because the collector-emitter saturation voltage is higher in silicon transistors.)

This circuit is unusual in that no feedback windings are used. Rather start-up regeneration results from the cross-connected 100-ohm resistances. After start-up, oscillation is largely governed by the saturation characteristics of the transformer core. That is, when the core saturates because of heavy conduction in one transistor, the transistor continues to conduct until the energy stored in the core is used up, whereupon that transistor abruptly turns off and the alternate transistor turns on. The buildup and collapse of the magnetic flux in the core alternately reverse direction, thereby timing the switching cycles. (An air-core version of such a circuit would also oscillate but in a mode more suggestive of a multivibrator—the L/R ratio of the windings, rather than magnetic saturation phenomena, would then govern the oscillation frequency.)

The output filter, consisting of L1, C1, and C2, "de-spikes" the output waveform and also modifies its shape from square to trapezoidal. Such a waveshape is less likely to produce "buzzing" in electronic equipment such as recorders, radios, or phonographs. If the inverter is intended to be used only for electric shavers, as if often the case, the filter network can be eliminated. For the sake of convenience it is suggested that the inverter be equipped with a plug for the cigarette lighter and a conventional ac outlet receptacle.

A PROTOTYPE 60-HERTZ INVERTER

The inverter shown in Fig. 4-19 is a classic in more than one sense. It is an early example of applied power electronics but remains the basic arrangement for numerous inverters still being made by hobbyists and electronic manufacturers. The 2N278 germanium transistors were originally employed in this circuit, but more modern germanium power transistors may be used, such as the Motorola 2N2081 or 2N1558. Despite its design vintage, this inverter has several features that are not readily excelled even with more modern semiconductor devices. The circuit is a typical example of the single-transformer saturable-core oscillator. This configuration, with its electromagnetic feedback, is superior to the more primitive circuit of Fig. 4-18. Its frequency is more nearly a function of readily manipulated factors than is the case with the cross-coupled resistor feedback technique. Higher power levels can be attained by scaling up core dimensions and wire sizes as well as transistor types or heat removal. The grounded collector circuit enables metal-to-metal contact between the transistor cases and the heat sinks.

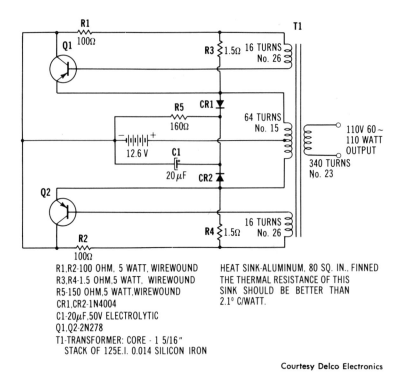

R1,R2-100 OHM, 5 WATT, WIREWOUND
R3,R4-1.5 OHM,5 WATT, WIREWOUND
R5-150 OHM,5 WATT,WIREWOUND
CR1,CR2-1N4004
C1-20μF,50V ELECTROLYTIC
Q1,Q2-2N278
T1-TRANSFORMER: CORE - 1 5/16"
    STACK OF 125E.I. 0.014 SILICON IRON

HEAT SINK-ALUMINUM, 80 SQ. IN., FINNED
THE THERMAL RESISTANCE OF THIS
SINK SHOULD BE BETTER THAN
2.1° C/WATT.

Courtesy Delco Electronics

Fig. 4-19.   A prototype circuit for inverters.

The use of standard silicon E-I transformer laminations is a blessing to the home experimenter who does not have a toroidal coil winder.

Inasmuch as this inverter is presented as a reference circuit for inverters with various performance parameters, the following equations are given to facilitate design:

$$f = \frac{V \times 10^8}{4B_s \times A \times N}$$

where

f is the frequency in hertz
V is the voltage across one-half of the primary (center-tapped) winding
B_s is the saturated flux density of the core in lines/cm²
A is the cross section of the core in cm²
N is one-half the number of turns of the total primary winding

Generally the saturated flux density will be obtained from specifications pertaining to the type of core material used. Voltage V is
approximately the battery voltage, but more accurately the battery voltage minus the collector-emitter voltage of one transistor (V_{CEsat}). For
germanium power transistors this will be on the order of 0.3 volt. A
typical value for silicon transistors is 1 volt.

Note that inasmuch as V, A, B_s, and f are determined from performance requirements, tables, or necessary guesstimates, the usual
procedure will involve solving for N. Accordingly the following per-
mutation of the fundamental equation will be found useful:

$$N = \frac{V \times 10^8}{4B_s \times A \times f}$$

The number of turns on the secondary winding is determined as follows:

$$N_s = \frac{V_s \times N}{E}$$

where

N_s is the number of turns on secondary winding

N is one-half the number of turns of total primary winding

V_s is the voltage across secondary winding

E is the voltage applied to one-half of primary winding. (This
is the battery or supply voltage minus the collector-emitter
voltage drop of one "on" transistor.)

The number of turns on the feedback windings is often made to
develop 5 or 6 volts rms. This permits the use of current-limiting base
resistances, such as R1 and R2 in Fig. 4-19, and is ample to produce
reliable starting. Computation of the number of turns for the feedback
winding is performed exactly as for the secondary winding.

Wire size is often selected on the basis of 750 milliamperes of
current per circular mill of wire area. Standard wire tables list the cross-
sectional area of the various wire gauges in circular mills. Experience
has shown that such current density tends to be both safe and economical. However, the cited design equations neglect the effect of wire
resistance on operation. Thus, because of resistance in the primary
winding, the *effective* voltage acting across N turns of the primary winding will be less than E. Also the voltage available from the secondary

winding will drop as soon as current is drawn through the secondary resistance. Because of these factors it is customary to overwind the secondary winding by 5 percent or so. These considerations were omitted from the design equations for the sake of simplicity.

Close coupling between the primary and the feedback windings is important in order to minimize the amplitude of voltage spikes that are the consequence of leakage inductance. One approach is to wind the primary winding directly over the core insulation and then distribute the feedback windings over the entire length of the primary winding. Bifilar winding techniques are often used to promote tight coupling. Because leakage inductance from less than unity coupling between the primary and secondary windings also contributes to generation of destructive voltage spikes, the actual construction of the transformer necessarily entails conflicts and compromises.

The experimenter can readily adapt this basic inverter for use with larger germanium transistors, such as the MP505, or with silicon types, including Darlingtons.

SELECTION OF THE INVERTER VOLTAGE

In most technical literature dealing with square-wave inverters, a nominal output of 115 or 120 volts is indicated when these inverters are intended as substitutes for the 60-Hz power line. It stands to reason that some discussion is in order inasmuch as we are comparing sine waves and square waves. To begin with, let us confine our investigation to those situations where we wish to deliver power to electromagnetic devices, such as motors and relays.

When the square-wave output of an inverter is applied to an electromagnetic device, the inductance is usually sufficient to act as a fairly effective filter choke. This implies ever-increasing attenuation to the progressively higher odd harmonics that are the building blocks of the square wave. For practical purposes, the first harmonic, or fundamental, will be left to develop the electromagnetic torque in a motor or ac relay. In determining the amplitude of square-wave voltage that will exert the same effect on a motor as the rated sine-wave voltage, we must keep the following facts in mind:

1. The fundamental frequency of a symmetrical square wave is the frequency of a sine wave with the same repetition rate as the square wave.
2. The peak amplitude of the fundamental frequency in such a square wave is $4/\pi$ times the peak amplitude of the square wave. (The peak of the sine wave fundamental is actually *greater* than that of the square wave. This fact is derived from the Fourier theorem of harmonic analysis.)
3. In a square wave the peak and rms values of amplitude are identical.
4. In a sine wave peak amplitude is 1.41 times the rms value.

If we put these facts together, it turns out that the rms (or peak) value of the square wave *needed* to do the same work in a motor or relay as the rms sine-wave voltage of these devices is given by the following equation:

$$V_{sq} = \frac{1.41 \times \pi \times V_s}{4}$$

where

V_{sq} is the *required* rms (or peak) amplitude of the square wave
V_s is the *rated* rms voltage of the electromagnetic device

The equation may be numerically simplified as $V_{sq} = 1.11 \times V_s$.

Example: What should be the voltage output of a square-wave inverter intended to power a motor that is rated at 117 volts from the 60-Hz line? *Answer:* The inverter output voltage should be 1.11 × 117, or 129.9 volts.

In actual practice a little experimentation may be in order because the calculated equivalence between the applied square wave and the rated sine-wave power may not prove valid in all cases. What often happens is that motor *current* (which is responsible for development of electromagnetic torque) may be a reasonably good sine wave, but motor *voltage* will, at best, be a trapezoidal wave. The implication of this is that much of the harmonic energy of the original square wave remains available to produce eddy-current and hysteresis losses in the motor. Thus, an inverter-powered motor may run hotter than it would from the 60-Hz line. These harmonics may also degrade the torque capability of the motor. Often it is necessary to put less than rated power into the motor and to expect less than rated mechanical output from it. However, much depends on both the inverter and the motor.

The operation of electronic equipment from the square-wave output of inverters may tend to heat up power or isolation transformers because of eddy-current and hysteresis losses. However, unless the transformers are skimpily designed, this is not likely to be a problem. The successful use of the inverter may depend on whether the filter capacitors in the power supply of the inverter-powered equipment operate near the peak voltage of the rectified ac. If so, the use of an inverter having the *same* rms voltage as that of the nominal 60-Hz line may prove unsatisfactory because the internal dc voltage within the equipment may then be too low. Here a 10-percent boost in inverter output voltage may overcome the problem. Actually this problem is not as acute as one might initially predict, because the square wave is more efficient at keeping filter capacitors charged than is a rectified sine wave. There also tends to be less background "hum" in inverter-powered electronic equipment. On the other hand much trouble is frequently encountered from "hash" and rfi.

A HIGH-POWER INVERTER

Although simple 60-Hz inverters using germanium transistors have their obviously useful application area, different design approaches are needed in order to produce higher output levels economically. It is generally necessary to utilize higher dc input voltages, to employ silicon transistors, and to operate at higher frequencies. It is true that such an inverter may no longer constitute a power source for equipment intended for operation from the 60-Hz power line. However, an even greater application area is the use of inverters with subsequent rectification so that dc-to-dc "transformation" takes place. The combination inverter and rectifier is known as a converter, or a dc-to-dc converter, and is often the heart of regulated dc power supplies. Here we are concerned with just the inverter portion of this important building block.

Another expedient often used when high output is desired is to employ *two* transformers. The circuit remains a self-excited saturable-core oscillator, but the output transformer is designed so that it operates in its *linear* region, as do conventional transformers used in most electronic applications. A much smaller saturable transformer is connected in the input circuit of the inverter so that basically the same magnetic switching phenomenon occurs as in the single-transformer inverter. However, the efficiency of the two-transformer circuit is superior to that of the simpler arrangement because transformer losses are greatly reduced in its output circuit. (Allowing a transformer to saturate is a strict "no-no" to the power engineer who tries for efficiencies in the vicinity of 97 percent or so. If we must use core saturation, it is much better to have it occur in a small, rather than a large, transformer.) Another important feature of the two-transformer circuit is that its production of voltage spikes is minimal. This enables the transistors to work well within their SOA ratings.

Fig. 4-20A illustrates such a two-transformer inverter. This circuit is capable of delivering 400 watts to its load and also has good tolerance for the effects of inductive or capacitive loads. A separate starting circuit is used because the optimization of operating efficiency so vital in high-power equipment is not always compatible with other characteristics— in this case start-up ability (especially under load).

The nominal oscillation frequency is approximately 1 kHz. By changing the small saturable transformer, T1, the operating frequency can be varied over a wide range. For example, it is probable that different winding formats for T1 can provide oscillation from, say, 400 Hz to at least several kilohertz. This is because output transformer T2 merely matches the inverter to the load requirements. Unlike single-transformer inverters, the output transformer is *not* involved in frequency determination. Surprisingly some electronic equipment rated for 50- to 60-hertz operation will perform well from a 1,000-kHz source. However, this remains a relatively unexplored application area.

The trigger circuit shown in Fig. 4-20B supplies a pulse of forward bias to transistor Q2 in the inverter when switch Sw1 is placed in its "start" position. The magnitude and duration of this pulse are determined by resistor R_x and capacitor C_x. The values of these components are best arrived at by experimentation. Suggested initial values for the resistor and capacitor, respectively, are several hundred ohms and several tenths of a microfarad. The trigger circuit probably can be dispensed

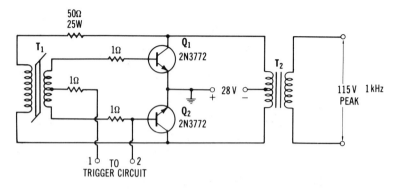

TRANSFORMER CHACTERISTICS

	T_1	T_2
CORE	SQ. ORTHONAL; MAGNETICS INC. No. 52035-2A	MICORSIL 150 E/U .004; 1.5 BY 1 INCHES
PRIMARY	67 TURNS No. 23 WIRE	20 TURNS No. 14 WIRE
SECONDARY	24 TURNS No. 18 WIRE	40 TURNS No. 18 WIRE

(A) Inverter circuit.

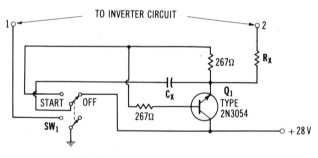

(B) Trigger circuit for initial start-up.

Courtesy RCA Corp.

Fig. 4-20.   Two-transformer inverter featuring high power and high efficiency.

with if too much demand is not made on the self-starting ability over a wide temperature range and with heavy or reactive loads. In any event it is not the two-transformer circuit per se that imposes the need of an external starting method. It is suggested that experimentation with the biasing circuit of the inverter may optimize self-starting ability if this is desired.

MOSFET POWER RF AMPLIFIERS

The advent of the power MOSFET introduced a new device into the domain of power electronics. Just as we are becoming acclimated to the use of these devices in such applications as audio-output stages, motor-control elements, and series-pass transistors for regulators, a new area of application as radio-frequency power amplifiers has evolved. Actually such use could have been anticipated, for the specifications of even the early Siliconix VMP-1 power MOSFET indicated that it could switch 1 ampere in about 5 nanoseconds. Other features suggesting its capability at radio frequencies were its high input and output impedance and its immunity to thermal runaway. Moreover, its operational resemblance to vacuum tubes, which have long served as Class-B and Class-C rf amplifiers was too compelling to be overlooked. It was only natural that the VMP-1 was soon followed by the VMP-4, a power MOSFET intended specifically for rf applications.

The MOSFET is ideal for rf power applications in vhf and uhf communications where a few watts, or tens of watts, into the antenna often suffice. At the lower radio frequencies, power output stages tend toward levels of hundreds and thousands of watts. Here too the MOSFET is becoming competitive.

Fig. 4-21A shows the circuit of a simple Class B (linear) amplifier that can deliver about 8 watts pep into a 50-ohm load at frequencies up to several hundred megahertz. The 16-watt pep push-pull version of this basic circuit is shown in Fig. 4-21B. Greater output power should be possible, particularly for interrupted service (cw transmitters). If the frequency is not too high, these MOSFETs can be paralleled easier than ordinarily is the case with bipolar transistors. Much enhanced electrical ruggedness is displayed by these devices compared to bipolar rf transistors. This is because the lack of secondary breakdown, together with a *positive* thermal coefficient, renders the MOSFET almost immune to the effects of load vswr and to the effects of detuning during experimentation and adjustment.

Some experimentation may be necessary with these circuits because of the drain capacitance introduced by the heat sinks. The use of physically small copper heat sinks in conjunction with beryllium oxide insulators can help keep heat-sink capacitance down. All things considered, however, the conventional practices pertinent to high-frequency layout and construction will suffice to obtain good results with MOS-

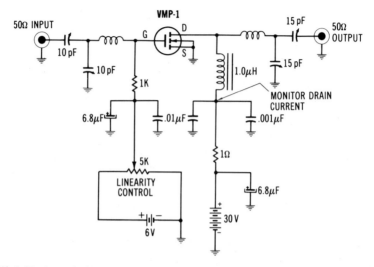

(A) A Single-ended linear amplifier with an output of approximately 5 watts pep.

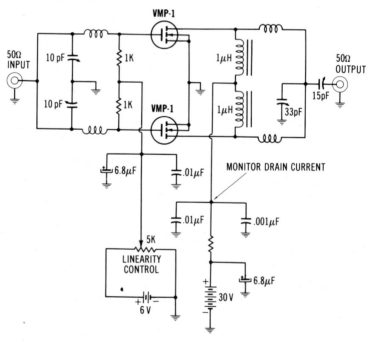

(B) Push-pull linear amplifier with an output of approximately 10 watts pep.

Courtesy Siliconix Inc.

Fig. 4-21. The power MOSFET used in rf amplifiers.

FETs. A small fraction of a watt of input power should prove adequate for Class B operation of these amplifiers. Somewhat greater drive will be needed with Class C operation. More advanced rf applications of power MOSFETs will be found in Chapter 6.

25-WATT VHF TRANSMITTER FOR MARINE
AND AMATEUR 2-METER BANDS

The enhanced reliability of marine communications brought about by progress in solid-state technology is truly phenomenal. Only a few years ago marine transmitters were 3-MHz am equipment and were predominantly vacuum-tube designs. While it is not necessarily valid to criticize tubes, low frequencies, or amplitude modulation, the combination of the three in boats left something to be desired. Whereas the construction of vhf gear is somewhat of an art as well as a science, experience and good engineering practices readily overcome the undesirable characteristics of high-frequency behavior. Even more of an art was the ability to make electrically short antennas absorb power efficiently from an otherwise satisfactory 3-MHz power stage. In many instances a good portion of the available power was dissipated in the losses of impedance-matching networks. Other problems included the fractional-ohm antenna impedances, the grounding difficulties, and rf "all over the place" except in the antenna.

The transmitter shown in Fig. 4-22 is intended to be driven from an fm exciter with a 100-milliwatt, 50-ohm output and with a carrier frequency in the range of 144 MHz to 175 MHz. Thus, operation is also feasible in the 2-meter amateur band and in land mobile radio band. Included in the rf circuit is a voltage-dropping circuit configured around the MJE2020 power transistor. This circuit is essentially a series-pass regulator with a bias adjustment (R2), which enables sufficient power supply voltage to be absorbed by the transistor so that the rf output of the transmitter can be reduced to 1 watt or less. Such a provision is stipulated by Federal Communications Commission (FCC) regulations. Otherwise the transmitter circuit is quite conventional. Resonant circuits are all lumped parameter designs, and "straight through" operation is intended within the specified frequency range. It should be appreciated that an rf power transistor such as the 2N6082 was available only to the military and space programs just a few years ago—and at a respectable

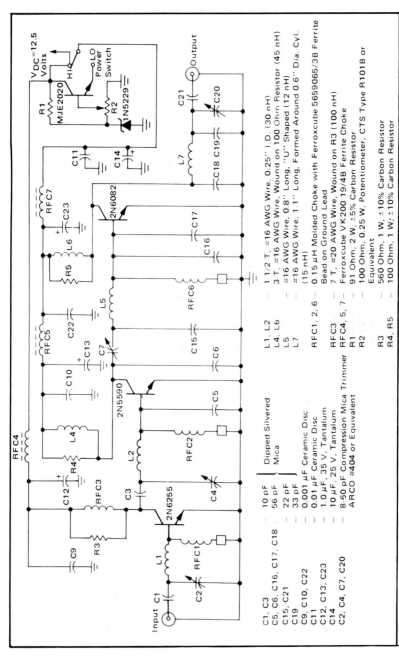

Fig. 4-22. Solid-state transmitter for the 144-MHz to 175-MHz vhf range.

Courtesy Motorola Semiconductor Products, Inc.

C1, C3 — 10 pF — Dipped Silvered
C5, C6, C16, C17, C18 — 56 pF — Mica
C15, C21 — 22 pF
C19 — 33 pF
C9, C10, C22 — 0.001 μF Ceramic Disc
C11 — 0.01 μF Ceramic Disc
C12, C13, C23 — 1.0 μF, 35 V, Tantalum
C14 — 10 μF, 25 V, Tantalum
C2, C4, C7, C20 — 8-60 pF Compression Mica Trimmer ARCO #404 or Equivalent

L1, L2 — 1-1/2 T, #16 AWG Wire, 0.25'' I.D. (30 nH)
L4, L6 — 3 T, #16 AWG Wire, Wound on 100 Ohm Resistor (45 nH)
L5 — #16 AWG Wire, 0.8'' Long, "U" Shaped (12 nH)
L7 — #16 AWG Wire, 1.1'' Long, Formed Around 0.6'' Dia. Cyl. (15 nH)
RFC1, 2, 6 — 0.15 μH Molded Choke with Ferroxcube 5659065/3B Ferrite Bead on Ground Lead
RFC3 — 7 T, #20 AWG Wire, Wound on R3 (100 nH)
RFC4, 5, 7 — Ferroxcube VK200 19/4B Ferrite Choke
R1 — 91 Ohm, 2 W, ±5% Carbon Resistor
R2 — 100 Ohm, 0.25 W, Potentiometer, CTS Type R101B or Equivalent
R3 — 560 Ohm, 1 W, ±10% Carbon Resistor
R4, R5 — 100 Ohm, 1 W, ±10% Carbon Resistor

price. An inexpensive device capable of 25 watts output up to 175 MHz was simply nonexistent until quite recently.

When an adequate heat sink is used, this output transistor has sufficient electrical ruggedness to be self-protected against the worst conditions likely to be encountered during tune-up, experimentation, or operation. This includes high battery voltage (15.5 volts) *together* with either a shorted or open-circuited load or load impedances anywhere between these extremes. However, such abuse can be endured only for short intervals. This is reminiscent of allowable overload periods with certain vacuum tubes.

A 200-WATT AUDIO AMPLIFIER
FOR STEREO SYSTEMS

Six npn silicon power transistors are utilized in the output stage of each amplifier channel in this high-power stereo system. In the circuit shown in Fig. 4-23, it will be seen that the push-pull output stage makes use of two groups of three parallel-connected RCA 1B05 power transistors. The output stage circuit is so-called quasi-complementary-symmetry configuration. Although all the output transistors are npn types, the circuit operates *as if* the lower group of paralleled 1B05 transistors were pnp types. By such simulation, true push-pull performance is obtained and direct coupling to the speaker is feasible. Such direct coupling is beneficial to low-frequency response and dispenses with the practical difficulties of high currents through a large electrolytic coupling capacitor.

The quasi-complementary-symmetry circuit is particularly desirable when a rugged and reliable transistor is available in only one polarity. It has often been a problem to devise true complementary-symmetry output stages from a diverse selection of npn and pnp power transistors. With silicon transistors the pnp type has tended to lag behind the power capabilities of npn versions. There also remains the problem of matching electrical and thermal characteristics of complementary transistors. In any event the quasi-complementary-symmetry circuit has enjoyed widespread use and is capable of excellent performance.

It will be seen that the immediate driver stages for the output stage are also 1B05 transistors. These drivers operate as Class B amplifiers. Going back one more stage, the lower-level drivers are 1E02 and 1E03

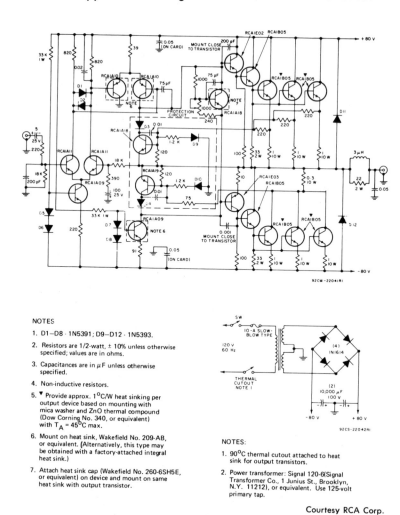

NOTES

1. D1–D8 · 1N5391; D9–D12 · 1N5393.

2. Resistors are 1/2-watt, ± 10% unless otherwise specified; values are in ohms.

3. Capacitances are in μF unless otherwise specified.

4. Non-inductive resistors.

5. ▼ Provide approx. 1°C/W heat sinking per output device based on mounting with mica washer and ZnO thermal compound (Dow Corning No. 340, or equivalent) with $T_A = 45$°C max.

6. Mount on heat sink, Wakefield No. 209-AB, or equivalent. (Alternatively, this type may be obtained with a factory-attached integral heat sink.)

7. Attach heat sink cap (Wakefield No. 260-6SH5E, or equivalent) on device and mount on same heat sink with output transistor.

NOTES:

1. 90°C thermal cutout attached to heat sink for output transistors.

2. Power transformer: Signal 120-6(Signal Transformer Co., 1 Junius St., Brooklyn, N.Y. 11212), or equivalent. Use 125-volt primary tap.

Courtesy RCA Corp.

Fig. 4-23. 200-watt quasi-complementary-symmetry amplifier.

transistors, the first being an npn type and the second being a pnp type. It is by virtue to these two opposite-polarity driver stages that the output stage is made to operate as if it were an npn-pnp combination.

Completing the cascade are two more driver stages, each operating as a Class A amplifier. These are the differential stage consisting of the pair of RCA 1A10 transistors, and finally the input differential stage, which includes a pair of RCA 1A11 transistors. The differential input stage provides a convenient method for injecting negative feedback from the output stage. Both of these differential stages are operated from constant-current sources in the form of individual circuits involving RCA 1A09 transistors. (When a differential amplifier is operated from a constant-current source, its common-mode rejection and dc stability are enhanced.)

In a high-power amplifier the biasing of the output transistor is an important consideration. True Class B operation is not desirable because of crossover distortion when the transistors switch conduction states. A small amount of forward biasing alleviates this situation but also degrades the operating efficiency. It is desirable to maintain a small amount of forward biasing, but it should be no greater than is consistent with an acceptable reduction in crossover distortion. This does not occur without special provisions because both the leakage current and the bias requirements of transistors are temperature dependent. A biasing provision is needed to maintain the idling current of the output stage approximately constant over a wide temperature range.

This function is provided by the so-called V_{BE} multiplier stage configured around the RCA 1A18 transistor. The base-emitter voltage of this stage is the bias source for the Class B drivers and thus for the output transistors. The 1A18 transistor is mounted on the same heat sink with an output transistor. This constitutes a thermal feedback arrangement and provides sufficiently good temperature tracking to keep the idling current of the output stage nearly constant. The adjustment of the idling current is provided by the 1,000-ohm variable resistance in the base circuit of the 1A18 stage. This V_{BE} multiplier operates from one of the constant-current sources. Without this circuit provision for such a stage the base-emitter voltage would not be solely a function of temperature.

We have thus far accounted for all of the functions except that provided by the two transistors and associated components enclosed within the dashed lines. This is a protective circuit that ordinarily does not alter the operation of the amplifier, but it will act to limit the maximum current consumed by the output stage. The sensing of output-stage current is accomplished via the three 220-ohm resistors in the emitter circuits of one group of paralleled 1B05 power transistors and the 75-ohm resistor connected to the collectors of the other group of paralleled 1B05 power transistors. The control of output-stage current is then dependent on the conduction of the two transistors enclosed by

Measured at a line voltage of 120 V, T_A = 25°C, and a frequency of 1 kHz, unless otherwise specified.

Power:	
Rated power (8-Ω load, at rated distortion)	200 W
Typical power (4-Ω load)	300 W
Typical power (16-Ω load)	130 W
Total Harmonic Distortion:	
Rated distortion	0.5%
IM Distortion:	
10 dB below continuous power output at 60 Hz and 7 kHz (4:1)	0.2%
IHF Power Bandwidth:	
3 dB below rated continuous power at rated distortion	5 Hz to 35 kHz
Sensitivity:	
At continuous power output rating	900 mV
Hum and Noise:	
Below continuous power output:	
Input shorted	96 dB
Input open	84 dB
With 2 kΩ resistance on 20-ft. cable on input	94 dB
Input Resistance	18 kΩ

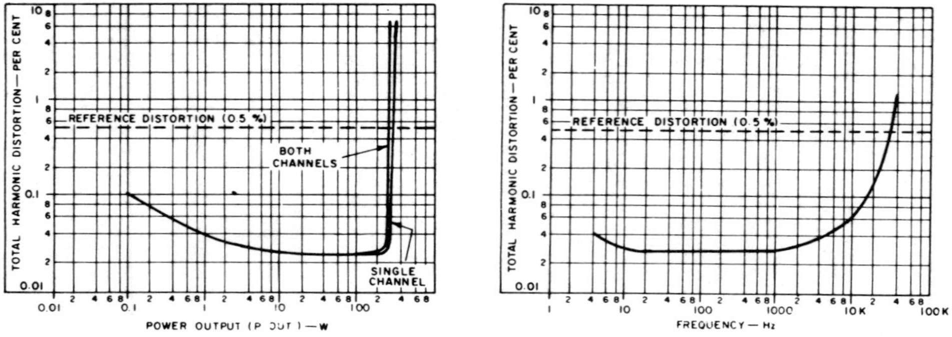

Fig. 4-24. Performance data for 200-watt quasi-complementary-symmetry amplifier. Courtesy RCA Corp.

the dashed lines. This is because these transistors are "active" loads for the second driver stage. When these loads increase conduction, less gain is developed in this driver stage and the drive to the subsequent stages is automatically reduced. Because of this self-regulating action, it is not possible to drive the output stage much beyond the rated output power of 200 watts. The performance data and curves pertaining to this amplifier are shown in Fig. 4-24.

100-WATT SERVO AMPLIFIER
FOR INDUSTRIAL APPLICATIONS

The direct-coupled amplifier shown in Fig. 4-25 incorporates the general design techniques used in audio and stereo equipment. However, here the emphasis is on simplicity, reliability, and electrical ruggedness. The high-voltage transistors employed in the quasi-complementary-symmetry output stage enable operation directly from rectified 120-volt ac line voltage. The DTS-430 npn transistors have maximum ratings of 400 volts for V_{CEO}, 5 amperes collector current, and 150 watts of power dissipation. They are made by a triple-diffusion process and have inordinately good second-breakdown capability. Consequently the once-valid notions that high-voltage operation necessarily entailed fragile reliability are strictly passe. Indeed, high voltage and relatively low current operation circumvent certain problems that all too often crop up in low-voltage, high-current operation. In order to instill further peace of mind regarding 150-volt operation of power transistors, it is well to realize that 1,500-volt transistors are commonly used in tv applications.

Output transistors Q7 and Q6 are driven by Darlington-connected drivers Q4 and Q5, respectively. Because Q5 is a pnp transistor, Q6 also operates as a pnp type even though it is actually an npn—thus the nomenclature *quasi*. Resistance R10 adjusts the transconductance of the Q5-Q6 pair to have the same transconductance as that of the Q4-Q7 combination. A further provision for bringing about near identical operation from both branches of the push-pull configuration is the injection of positive feedback into the base of Q5 via C3 and the network consisting of resistors R7 and R8. (The amplifier incorporates a *net* negative feedback of 40 dB, so this technique does not adversely affect stability. It merely helps attain symmetrical outputs from both branches of the amplifier.)

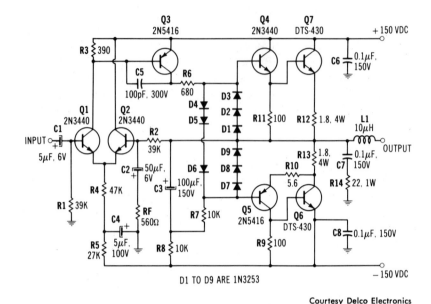

Fig. 4-25. 100-watt servo amplifier designed to operate directly from rectified ac line voltage.

The output stage develops its rated 100 watts into a 75-ohm load. This, of course, is considerably higher than the 4- and 8-ohm speaker impedances commonly encountered in audio work. Output transistors Q6 and Q7 operate in Class AB with about 20-mA idling current.

This amplifier develops a high open-loop voltage gain because Class A driver transistor Q3 "sees" a high impedance at the bases of the Darlington drivers, Q4 and Q5.

SERVO AMPLIFIER FOR TWO-PHASE INDUCTION MOTOR

For applications of servo systems to aircraft, radar, and instrumentation systems, the two-phase induction motor is often considered advantageous to the dc motor. The *motor* itself is preferable to a dc motor because there is no commutator. This dispenses with brush sparking and, of course, the maintenance inherent in any machine using brushes. Also, the two-phase induction motor is efficient and results in considerable reductions in size, weight, and cost, especially for 400-Hz or higher frequency systems. There are compelling advantages in the servo amplifier as well. For example, the fact that stages of the amplifier can be ac coupled circumvents problems pertaining to drift and zero-signal setting. In the event of unusually abusive conditions or a component failure, the ac-coupled amplifier is not as likely to suffer from the over all catastrophic destruction that occurs all too commonly with direct coupling.

Hitherto, some design difficulties have often ruled out the two-phase servo system. In circuits where the quadrature phase relationship has been established by the use of capacitors connected with the motor windings, these capacitors tended to be quite expensive because they must have high capacity and high current ratings. Electrolytic types, though apparently alleviating some problems, add others such as temperature sensitivity, high dissipation factor, aging, and limited life.

The servo amplifier for two-phase induction motors shown in Fig. 4-26 enables realization of the salient features inherent with the use of the induction motor without the problems that have previously plagued actual implementations. Because of the high-voltage power transistors in the output stage, it becomes feasible to use 115-volt motors. No capacitors are associated with the motor windings. Although the circuit shown is set up for 60-Hz systems, it is only necessary to apply the proper scaling factor to the phase-shift and coupling capacitors to adapt the circuit to 400-Hz or 800-Hz systems. (Multiply the values of these capacitors by 60/400, or 0.15, to convert to 400-Hz operation.)

The Darlington-connected transistors in the complementary-symmetry output stage operate essentially as a push-pull, Class B power amplifier. (A *slight* amount of forward bias is applied in order to reduce the dead zone.) When powering a size 18 motor, the output stage develops a voltage gain of 20 (26 dB) and the overall voltage gain of the amplifier is 1,600 (64 dB). Average power supply currents are 70 milliamperes. Transistors Q3 and Q5 require the Motorola MS-10 heat sinks (or equivalent) in order to hold transistor case temperature to 40°C with an ambient temperature of 25°C at full power output.

The driver stage is an MC1709 op amp. This stage is arranged to operate from a single dc supply and is connected as an inverting amplifier. Single-ended output suffices to drive the push-pull section through coupling capacitors C1 and C2. Thus, no transformers are needed in the amplifier circuit.

The function of the first stage is to provide the required phase shift so that the ac voltages impressed across the two motor windings will

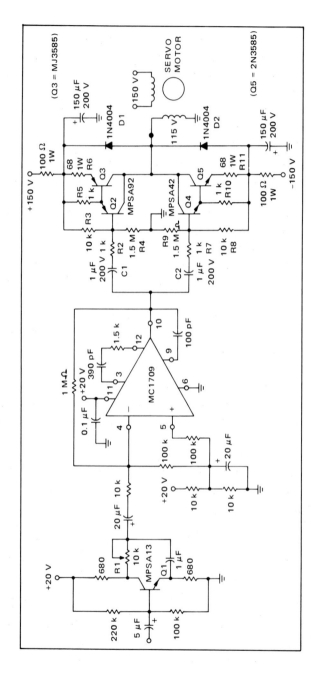

Fig. 4-26. Servo amplifier for two-phase induction motor.

Courtesy **Motorola Semiconductor Products,** Inc.

226

differ in phase by 90 degrees. To accomplish this, the first stage in the amplifier must produce an even greater phase shift than this in order to compensate for the *opposite* phase shifts produced by the interstage coupling capacitors. Potentiometer R1 provides an adjustment for attaining the exact quadrature phase relationship at the motor windings. Although not shown as such, the MPSA-13 is actually a monolithic Darlington transistor. Because of the equal collector and emitter resistances, this stage provides unity voltage gain. The trade-off obtained is high input impedance and exceptional stability. A unique feature of this type of "active" phase-shifting circuit is that, unlike a simple passive RC circuit, the phase shift can be varied both above and below 90 degrees with negligible variation in output voltage. A suitable power supply is shown in Fig. 4-27. No filtering is needed in the high-voltage section.

Two techniques for bringing about closed-loop positioning control of a bi-phase induction motor are shown in Fig. 4-28. Both schemes

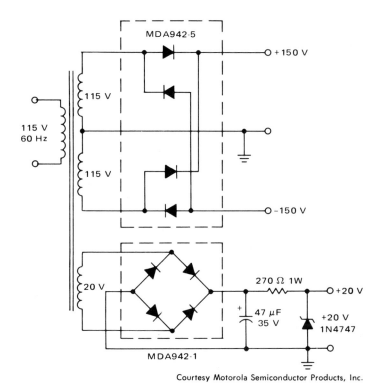

Fig. 4-27. Power supply for the induction-motor servo amplifier.

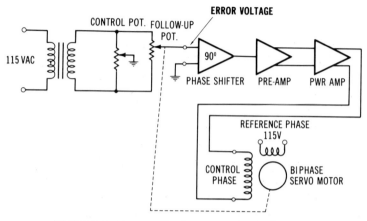

(A) Error signal generation by means of potentiometers.

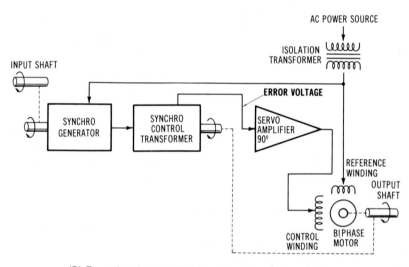

(B) Error signal generation by means by synchro devices.

Courtesy Motorola Semiconductor Products, Inc.

Fig. 4-28.   Closed-loop control methods using the servo amplifier and bi-phase motor.

provide follow-up sensing of the motor shaft position because of the tendency for the error signal at the input of the amplifier to null itself. At null, the control winding of the motor is deprived of excitation and no torque is developed. However, when there is an angular discrepancy between the positions of input and output shafts, the control winding is energized by a quadrature voltage. The greater the discrepancy, the more voltage is applied to the control winding and the greater is the corrective torque developed in the motor. (In Fig. 4-28B the three sections of the servo amplifier are lumped together.)

REGULATING SPEED CONTROLLER
FOR PERMANENT-MAGNET MOTOR

Before the advent of modern ceramic magnetic materials, the permanent-magnet dc motor was not extensively used. Such motors were actually larger than their shunt-wound counterparts and were subject to changes in the magnetization of their pole structures. In general, their operating characteristics offered no compelling features. However, it is another story with the newer permanent-magnet motors. They are physically small, their cost is reasonable, and their operation is superior to wound-field motors for many purposes. In particular they have a high starting torque, and their speed is a fairly predictable function of the torque exerted by the shaft when turning a mechanical load. This motor lends itself readily to electronic speed control. The circuit of Fig. 4-29 provides at least a 3-to-1 range of speed, and any selected speed is regulated from zero (no load) to fully rated output torque. The speed constancy exceeds that of squirrel-cage induction motors. However, by appropriate selection of circuit components and perhaps other minor modifications, the speed regulation can be even further improved.

The control and regulating features of this circuit are derived from a pulse-width modulation technique in which the pulse repetition rate is governed by optically generated pulses. The mechanism involved in producing these pulses utilizes reflected light from black-and-white segments painted on the motor armature. (Of course, a shaft-coupled cylinder or disk can also be used.) The initial conversion of pulsed light to pulsed current is made by photo transistor Q1. Each "package" of reflected light is equivalent to a pulse of forward base bias applied to a conventional transistor. The fiber optics are shown in simplified form— best results are likely to be obtained with a bundle of fibers conveying light from the lamp to the "chopper." Also some consideration will have to be given to the effect of ambient light on the painted surface of the armature.

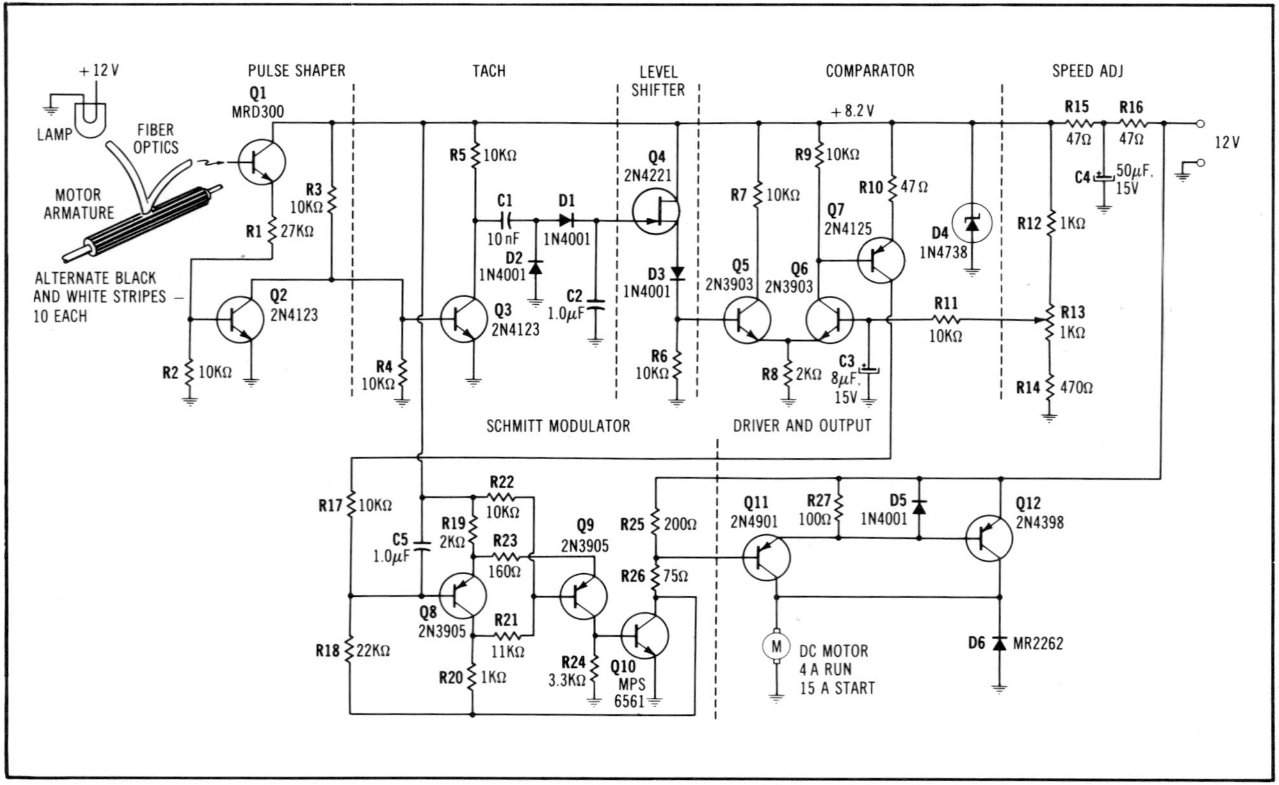

Fig. 4-29. Regulating speed-controller circuit for permanent-magnet motor.

Courtesy Motorola Semiconductor Products, Inc.

When considered on a functional basis, photo transistor Q1 and amplifier stage Q2 constitute a pulse-shaping circuit. The objective is to produce pulses of equal amplitude despite variations in the reflectivity of the white stripes. These variations are inevitable due to the non-uniform conditions of the optical surface. The effective load resistance of stage Q2 is quite high. Collector clipping readily occurs and all output pulses are thereby "sliced" to the same voltage level. Such a train of uniform-amplitude pulses enables the next section of the circuit to operate as a true tachometer, wherein the output voltage level is proportional to pulse repetition rate and therefore to motor speed.

The *tachometer* circuit involves transistor Q3 and associated components. In this circuit, Q3 provides polarity inversion of the pulse train. The diode-capacitor network in the collector circuit of Q3 constitutes a charge-pump circuit in which capacitor C2 develops a positive voltage proportional to the incoming pulse rate.

Field-effect transistor Q4 shifts the dc level of the tachometer to comply with the circuit needs of the next stage, which is a voltage comparator. Diode D3 serves to compensate the controller against changing ambient temperature.

In the *comparator*, transistors Q5 and Q6 function as a differential amplifier, and transistor Q7 is an output stage. *One* input of the differential amplifier samples the tachometer voltage, whereas the *other* input samples an adjustable "reference" voltage. The magnitude and polarity of the output voltage from the comparator depend on the difference or "error" between the sampled inputs. The generation of the error signal is of great importance in any feedback or servo system. In this controller the error signal is used to vary the duty cycle of current pulses delivered to the motor. This is accomplished via a Schmitt-trigger circuit, which functions as a pulse-width modulator.

The *Schmitt modulator* consists of transistors Q8, Q9, and Q10. Transistors Q8 and Q9 form the basis of the Schmitt-trigger circuit, and transistor Q10 operates as an output stage. The Schmitt-trigger circuit regeneratively latches into one or the other of its two conductive states, depending on the polarity of the error signal received from the comparator. In the overall scheme this enables the motor to *receive* current as long as its speed is too low and to be *deprived* of current as long as its speed is too high. This is what is implied by *pulse-width modulation* in this controller circuit.

The actual motor current is "metered" out by the Darlington arrangement comprising transistors Q11 and Q12 in the output circuit.

Diode D6 provides a motor-current path when the output stage is in its nonconductive state. This current is derived from the energy *stored* in the motor inductance during on time. The effect of this *free-wheeling* diode is to make the actual motor current *continuous* rather than pulsed. In a real sense, such a diode, in conjunction with the motor inductance, behaves as a *filter*.

The performance of this controller is illustrated by the curves in Fig. 4-30. It merits consideration in such applications as phonographs, recorders, and chart drives.

WIDE-RANGE VARIABLE SPEED CONTROL
FOR A 2-HORSEPOWER INDUCTION MOTOR

For many years the three-phase induction motor has been the workhorse for numerous industrial processes. Such application status is well deserved as these motors are efficient, relatively maintenance free, and

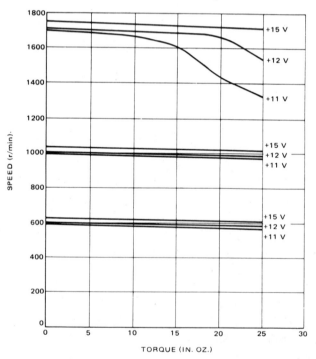

(A) Speed vs. torque for different supply voltages at approximately 40°C.

Fig. 4-30.   Performance curves for regulating speed-controller circuits.

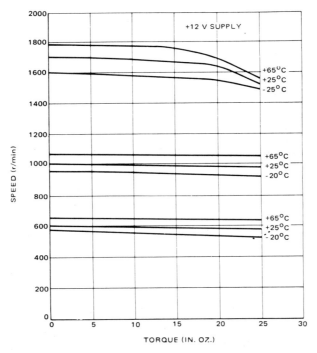

(B) Speed vs. torque at three different temperatures.

Courtesy Motorola Semiconductor Products, Inc.

Fig. 4-30. *continued*

readily reversed. An additional performance feature that has proved a mixed blessing is the nearly constant speed characteristics of these motors. At no load the shaft rotation approaches the synchronous speed of the rotating magnetic field; at full load the speed drops a nominal 5 percent or so, depending on design. For many applications such characteristics are extremely useful. However, for many *other* uses, the otherwise compelling features of the induction motor are to no avail because of its inability to provide other than small amounts of speed variation. Until recently the only suitable method of obtaining extensive speed control has been by means of a motor-alternator set. Obviously such an installation has a high initial cost. However, designers never lost sight of the simple fact that good speed control was forthcoming from a *variable-frequency power source*.

As may already be surmised, power electronics obligingly comes to the rescue. The equipment to be described here is essentially a variable-frequency inverter with three-phase outputs for speed control of a 2-horsepower, 480-volt, three-phase induction motor. The frequency can be manually adjusted over the range of 5 Hz to somewhat beyond 60 Hz, thereby enabling a 12-to-1 speed range to be attained (approximately 145 rpm to 1750 rpm). A practical feature of this inverter is that the voltage it delivers to the motor is directly proportional to the frequency, which is a requirement of induction motors.

Although the dc power supply is not described, it must be capable of supplying 7 amperes and its output voltage must be controllable from 0 to 750 volts. A single-phase design using a variac in the power line, an appropriate transformer, and a bridge rectifier should work well. However, a three-phase design with ganged variacs would be preferable from the standpoint of dc quality. A switching-type dc power supply, or some thyristor control arrangement, also merits consideration. In any event it is the *manual adjustment of the dc supply voltage* that varies the frequency of the inverter and, therefore, the speed of the motor.

The block diagram of the variable-frequency inverter is shown in Fig. 4-31. Some of its salient features follow.

1. Motor voltage is directly proportional to frequency.
2. Output transistors are operated in switched mode for optimum efficiency.
3. The three-phase voltage format for the motor is processed by CMOS digital logic circuits.
4. A six-segment waveform is synthesized nearly as good as a sine wave for the motor.
5. "Dead zones" are inserted in the waveforms so that overlapping conduction of output transistors cannot occur.
6. The inverter power section is separated by optoisolators from the logic section.
7. Current limiting circuitry protects the motor, the inverter, and the power supply from overloads.
8. Reverse rotation can be accomplished via appropriate switching at logic level.
9. ICs and Darlington transistors are advantageously used to minimize the parts count.
10. There are no commutation problems, as there are with thyristors.

Before dealing with the actual circuitry of the speed-control inverter, some basic insights can be acquired by taking a preliminary look

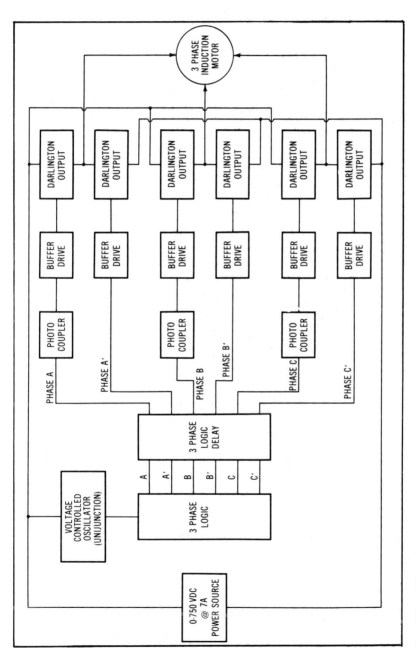

Fig. 4-31. Block diagram of variable-frequency inverter for speed control of three-phase induction motor. Courtesy Delco Electronics

235

at the waveform diagram in Fig. 4-32. The three lower waveforms represent idealized currents delivered to the stator windings of the three-phase motor. These waveforms appear to be a far cry from sinusoidal shape. A question might well arise with regard to the possibility of utilizing square waves, which are much easier to generate. However, this peculiar six-segmented wave possesses three desirable features. First, the third-harmonic component, which produces much of the hysteresis and eddy-current losses when square waves are used, is almost negligible in this unique waveform. Second, the filtering action of the motor inductance is sufficient to make the *actual* motor current smooth enough that the motor behaves nearly as well as with sine waves. (It is the current wave, rather than the voltage wave, that manifests itself as electromagnetic torque in the motor.) Third, the *dead zones* protect the output transistors from the destructive effects of overlapping conduction. Because of the three-phase format, the motor suffers little torque depletion due to these chopped waves. Such a motor is analogous to a

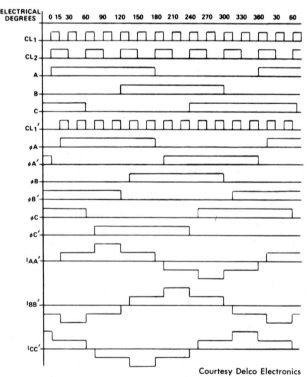

Courtesy Delco Electronics

Fig. 4-32. Logic timing diagram for variable-frequency speed-control inverter.

multicylinder automotive engine in which nearly constant torque is developed during rotation of the output shaft.

Now that we are mindful of the ultimate objectives of this system, let us inspect the circuit and see how the desired operation is brought about. The voltage-controlled oscillator and three-phase logic circuitry are shown in Fig. 4-33. (Only one such circuit is required for the system.) The voltage controlled oscillator is the 2N2646 unijunction stage. Whereas base 2 goes to a fixed 8-volt source, the emitter circuit receives its voltage from the 0–750-volt main supply. This enables the UJT to produce pulses with a repetition rate that increases and decreases proportionally to the applied dc voltage. In this way a 60-Hz, 480-volt motor performs satisfactorily down to a frequency of 5 Hz and a voltage of 40 volts.

All of the gates and flip-flops are CMOS logic. They exhibit high immunity to false response from the transients and noise pulses generally encountered in industrial environments. Integrated circuit IC13 operates essentially as a buffer, or driver, stage. Flip-flop IC1 then divides the UJT rate by two and provides the clock-signal pulses on which all subsequent waves are based. This important pulse train is designated CL_1 in the logic timing diagram (Fig. 4-32) and in the logic circuitry of Fig. 4-33.

The IC9 integrated circuit is a flip-flop that performs another division by two, thereby providing pulse train CL_2, which is another building block for later fabrication of the stepped waveforms delivered to the motor. The important three-phase voltage format first appears at circuit points A, B, and C, along with the complement waveforms from circuit points A', B', and C'. All of these waveforms are the consequence of a 3-bit shift register formed by flip-flops IC10, IC11, and IC12.

Both NAND gates IC15 and IC16 are necessary to prevent the shift register from performing in a forbidden mode in which the flip-flops would toggle with every clock pulse. Finally, NAND gate IC14 produces clock pulse CL_1'. The special purpose of this clock pulse is to impose a 15-degree deadzone in the current waveforms delivered to the motor. As mentioned, this technique ensures that no overlap of conduction can occur in the output transistors.

Referring to the schematic diagram of the inverter, Fig. 4-34, it will be observed that the incoming signals are processed by additional digital circuitry. (This inverter is one of three identical units required to drive the motor.) Signals A and A' are applied to the D inputs of D flip-flops IC5 and IC2, respectively. These flip-flops are clocked by the

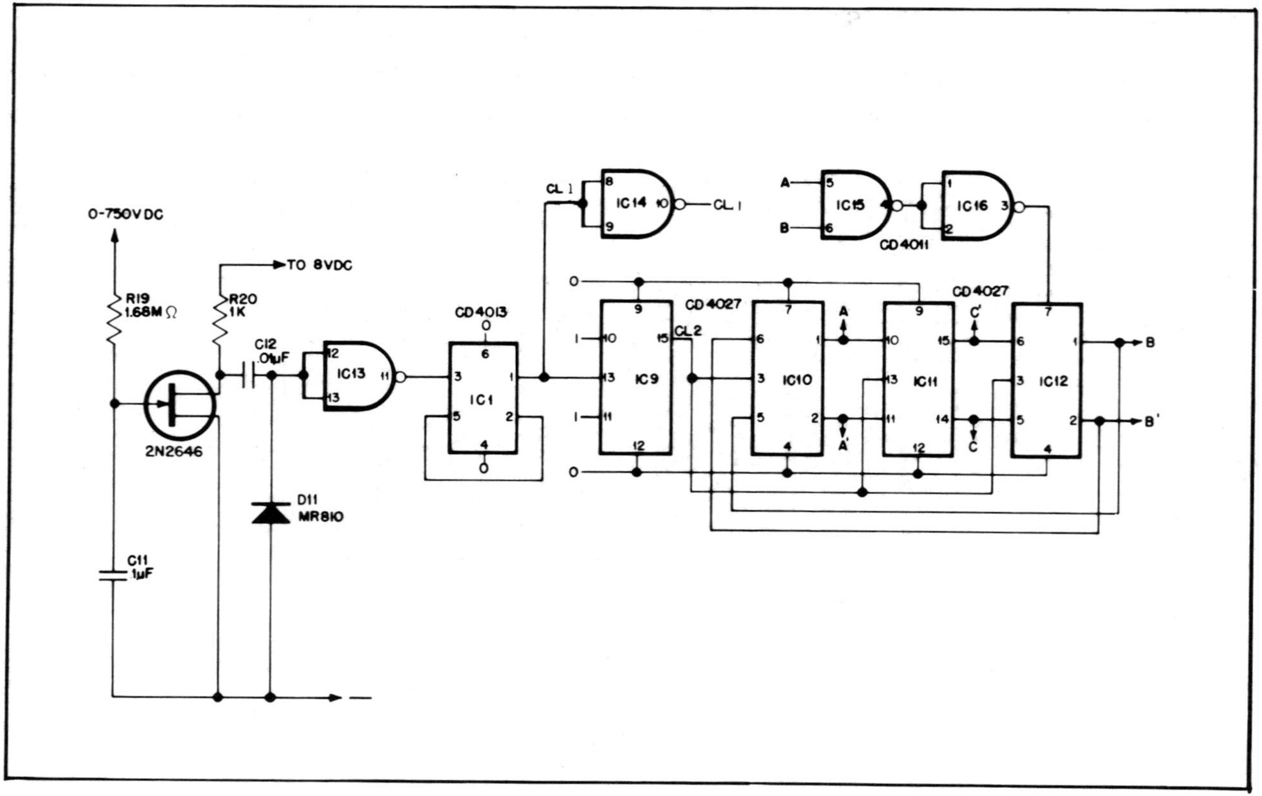

Fig. 4-33. Voltage-controlled oscillator and basic three-phase logic for speed-control inverter. Courtesy Delco Electronics

238

CL_1' pulses. The waveform diagram reveals that such clocking produces 15-degree deadzones in the output waveforms. Both IC4 and IC6 provide polarity inversions needed for proper drive of the inverter preamplifier stages. The IC3 integrated circuit is associated with over-current protection to be described later.

Each inverter unit has an optoisolator, such as PC1 in Fig. 4-34. This provides electrical isolation between the positive high-voltage channel and the logic and control circuitry. Darlington stage Q2 functions as a driver for the LED within the optoisolator. Each channel has DTS2000 Darlington preamplifier stages. These stages consist of Q3 and Q4 in the top channel and Q5, Q6, and Q7 in the lower channel. The additional stage (Q5) in the lower channel closely simulates the circuit effects of the optoisolator and its driver stage, Q2, on the upper channel.

The output stages of this inverter involve pairs of DTS709 transistors. The modified Darlington connection of these output transistors enables faster and more efficient switching to be accomplished than with ordinary Darlington arrangements. Basically the technique involves the application of negative bias to the base of the output transistors. Because of inductance in the motor, freewheeling diodes D5 and D9 are necessary, as are the snubber networks, such as R9, C1, and D4. These elements help maintain motor current and keep the operation of the output transistors within their SOA ratings.

Operating voltage for the logic circuit is provided by DVR-8, a three-terminal voltage regulator. Operating voltages for the inverter preamplifier stages are provided by the two center-tapped bridge supplies involving transformers T1 and T2. These supplies also provide the negative base bias for the output transistors.

Current-overload protection occurs through the deployment of IC3, IC7, and IC8. A NAND gate, IC8, senses the voltage drop across R18, which is a small resistance connected in series with the negative high-voltage lead. When sufficient overload current exists, IC8 inverts the sensed voltage drop and clears flip-flop IC7. In turn, base drive to preamplifier stage Q5 is removed because of the switched state of NAND gate IC3.

Actually the motor current interruption is not constant. Rather it endures only momentarily until the next phase transition occurs. At that time flip-flop IC7 is reset. If the fault has been eliminated, normal current is again delivered to the motor. Conversely, if the overload persists, the voltage drop across R18 will again trigger interruption of the motor

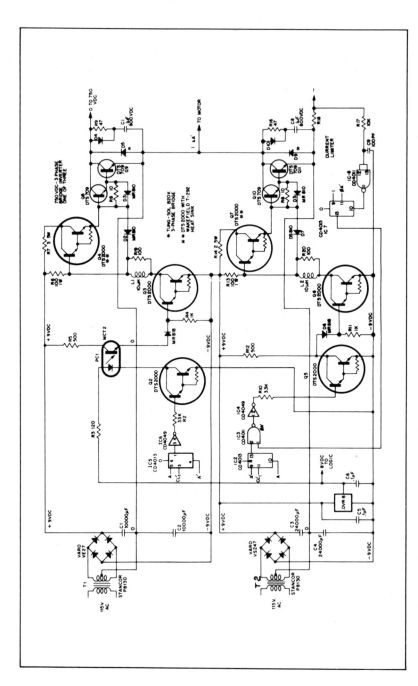

Courtesy Delco Electronics

Fig. 4-34. Schematic diagram for speed-control inverter.

current. This on-off cycling of the current-limiting circuit can prevail indefinitely with safe operation of the dc power supply, the inverter, and the motor. And as soon as the overload is removed, normal motor operation automatically resumes.

This project obviously is not intended for the inexperienced hobbyist or the neophyte experimenter. However, for the advanced amateur, the veteran technician, or the electronics engineer these circuit techniques will be suitable for many other applications. For example, adaptations can be made for smaller motors and even single-phase induction motors where simpler and less costly power supplies will be involved. On the other hand the power-oriented engineer will find it feasible to scale up the design to handle motors up to 10 horsepower. This could be accomplished by paralleling output transistors and providing for an even larger power supply than is needed for the described 2-horsepower system. Delco Electronics can supply such information in application note 58. Such speed controls have had revolutionary implications in medium-heavy industrial processes.

WARNING

The voltage and power levels involved in this motor-control system are potentially lethal. Only those who have had relevant experience with high-voltage and high-power circuits should undertake projects of this kind.

THE HYBRID-CIRCUIT POWER MODULE

Another device making a successful bid for a respectable slice of the linear power pie is the *hybrid-circuit module*. Contrary to some prevalent notions this circuit fabrication scheme did not become extinct with the advent of monolithic integration. Indeed modern hybrid-circuit modules not only make use of thick- and thin-film techniques and miniature printed-circuit ceramics but also utilize monolithic chips advantageously from both electrical and thermal considerations. The hybrid-circuit modules on the market appear to possess superior power ratings to monolithic ICs (excluding the monolithic Darlington, which is not generally classified as an IC).

Fig. 4-35 depicts a general-purpose amplifier employing the HC2500 hybrid circuit. In essence we have a high-power op amp with a 7-ampere

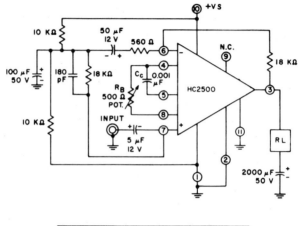

VS	54 V
P out	60 W
Idling Current (R$_B$ = 168 Ω	50 mA
THD	0.15%
IMD @ 50 mW	0.06%
V offset Pin 3 To Gnd.	+ 100 mV
Efficiency	64%
R L	4 Ω

Courtesy RCA Corp.

Fig. 4-35. General-purpose power amplifier utilizing the HC2500 hybrid-circuit module.

peak current rating. The basic arrangement shown can be readily adapted for service in a number of applications, such as audio, ultrasonics, servo systems, solenoid drivers, driven inverters, magnetic deflection of cathode-ray tubes, regulators, and motor control. From Fig. 4-36 we see that the internal circuitry of the HC2500 is as complex as the hi-fi and servo amplifiers discussed earlier in this chapter. Yet the complete circuit is contained in a small hermetically sealed metal case with a volume of less than 3 cubic inches. Exclusive of the mounting ears, the case is approximately 2 inches square and incorporates an electrically isolated

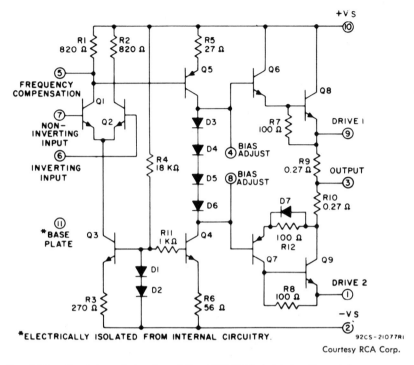

Fig. 4-36.   The internal circuitry of the HC2500 hybrid-circuit module.

base plate for making the required thermal contact with a heavy chassis
or a heat sink.

The device can be operated from either a single power supply or
a split power supply. Its bandwidth is specified at 30 kHz when delivering
60 watts. When operated with split 37.5-volt power supplies and adequate heat sinking, 100 watts output is feasible. This giant op amp has
a high open-loop gain (approximately 70 dB), but the feedback network
in Fig. 4-35 produces an actual circuit voltage gain of about 33.

A similar hybrid-circuit power amplifier made by RCA is the
HC2000H. This module includes a load-line limiting network that protects against high-energy transients from inductive loads. Also included
within the package are free-wheeling diodes for recovery of energy
stored in inductive loads, such as motors, during Class B operation.

As might be suspected, hybrid power modules can provide a cost-
effective blend of some of the salient features of *both* discrete and

monolithic devices. Because of the simplicity of implementing them into systems, they have a competitive edge over discrete devices. Also they merit consideration at power levels where true monolithic devices are not available. Thus hybrid power devices are likely to proliferate rather than diminish in popularity. This is especially true in light of the savings in parts-count that can be effected. This can be appreciated by comparing Fig. 4-36 with any of the stereo-amplifier circuits shown in this book.

5

Applications
Using Thyristors

UNIVERSAL POWER CONTROLLER

In addition to versatility of application, the phase-control circuit in Fig. 5-1 has a number of other features. Because the full-wave operation of a triac is utilized, the control range extends from nearly zero power to almost the maximum power that the ac power line can deliver to the load. Full-wave control enables the use of an isolation transformer, if so desired, with freedom from the core-saturating effect of the dc component inherent in half-wave control systems. Another advantage of full-wave phase control over half-wave circuits is the improvement in the power factor of the ac line, which enhances operating efficiency. Although these facts have been appreciated ever since thyristors became commercially available for the control of power, recent improvements in triac ratings have made the implementation of full-wave phase-control feasible at power levels previously the domain of the SCR. Although full-wave SCR circuits have been used with oppositely connected SCRs in parallel arrangements, the circuitry and operation are not always as simple or as economical as would be desired.

The circuit of Fig. 5-1, unlike many low-cost light dimmers, exhibits almost negligible hysteresis. Hysteresis manifests itself as differences in the setting of the control potentiometer at which the lamp turns on and turns off. It is a nuisance, not only because of the sloppiness of control,

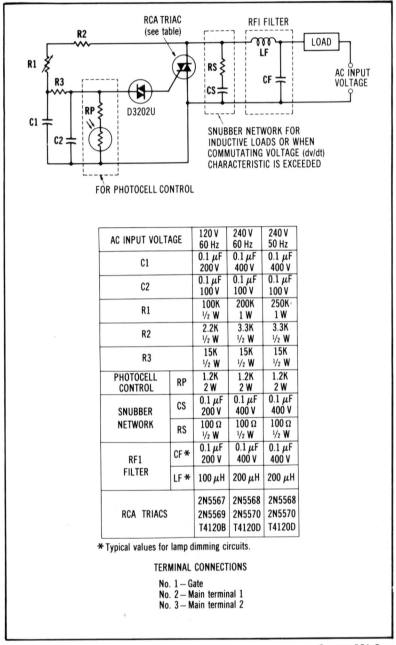

AC INPUT VOLTAGE		120 V 60 Hz	240 V 60 Hz	240 V 50 Hz
C1		0.1 μF 200 V	0.1 μF 400 V	0.1 μF 400 V
C2		0.1 μF 100 V	0.1 μF 100 V	0.1 μF 100 V
R1		100K ½ W	200K 1 W	250K 1 W
R2		2.2K ½ W	3.3K ½ W	3.3K ½ W
R3		15K ½ W	15K ½ W	15K ½ W
PHOTOCELL CONTROL	RP	1.2K 2 W	1.2K 2 W	1.2K 2 W
SNUBBER NETWORK	CS	0.1 μF 200 V	0.1 μF 400 V	0.1 μF 400 V
	RS	100 Ω ½ W	100 Ω ½ W	100 Ω ½ W
RF1 FILTER	CF *	0.1 μF 200 V	0.1 μF 400 V	0.1 μF 400 V
	LF *	100 μH	200 μH	200 μH
RCA TRIACS		2N5567 2N5569 T4120B	2N5568 2N5570 T4120D	2N5568 2N5570 T4120D

*Typical values for lamp dimming circuits.

TERMINAL CONNECTIONS

No. 1 — Gate
No. 2 — Main terminal 1
No. 3 — Main terminal 2

Fig. 5-1. Universal triac power controller.

but because a momentary drop in line voltage can cause the lamp to turn off at low light levels. If such an event occurs, turn-on can be restored only by manually advancing the control potentiometer. Hysteresis is caused by an abrupt drop in the gate-capacitor voltage when triggering occurs. This effect is greatly reduced by the use of a double-time-constant control circuit involving two gate capacitors. The double-time-constant circuit in Fig. 5-1 consists of resistances R1, R2, and R3 in conjunction with capacitors C1 and C2. Also contributing to preciseness of control is the diac gate-triggering device, which displays the symmetrical breakdown characteristics shown in Fig. 5-2. The use of the diac also makes triggering less temperature sensitive and makes triacs with widely different triggering characteristics behave approximately the same in the circuit.

The turn-on of a power thyristor such as the triac is abrupt and is accompanied by harmonics and electrical noise extending well into the radio-frequency region of the spectrum. These higher frequencies tend to back up into the ac line, which can cause interference to other electronic equipment. The transfer of such interference takes place by both conduction through the power line and by radiation from it. An rfi filter, such as included in the circuit of Fig. 5-1, greatly minimizes this phenomenon. The rfi filter is composed of a low-pass L section involving inductor L_F and capacitor C_F. (The filter also attenuates incoming high frequencies from the power line, which in some instances can produce erratic triggering.) The cut-off frequency of this simple filter is in the vicinity of 100 kHz.

When inductive loads such as motors are used, the commutation of the triac still tends to occur when the current goes through zero.

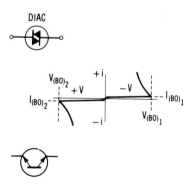

Fig. 5-2. Voltage-current characteristics for the diac trigger devices used with thyristors.

Courtesy General Electric Co.

However, the load current and load voltage are not in phase, as in the case with resistive loads such as lamps and heaters. This results in a high rate of voltage change across the main terminals of the triac. If such a voltage change is sufficiently high, it can cause enough current to pass through internal junction capacitance to behave as a gate trigger pulse. The triac would then be turned on even though an actual gate signal is not delivered from the triggering circuit at that time. Such false triggering not only disturbs the power control function of the circuit but also can be damaging to both the triac and the load. A common result of false triggering is latch-up, which is the failure to turn off at zero current. False triggering due to high rates of voltage change can occur in other ways, such as from switching at rates higher than 60 Hz or from voltage transients originating in other devices or circuits. *Snubber* networks are often employed to slow rates of voltage change and thereby eliminate false triggering. The snubber network used in the power control circuit of Fig. 5-1 consists of capacitor C_S and resistor R_S. This network, in conjunction with the anticipated load inductance, forms a critically damped LCR circuit. Ringing and overshoot of transient energy are largely absorbed by the resistive element of the snubber network.

Surprisingly the power to the load can also be controlled by a photocell inserted as indicated in Fig. 5-1. The overall power amplification involved in such control is tremendous and usually requires a number of cascaded amplifier stages in "linear" circuits. As revealed by the symbol, the photocell is of the photoconductive type, rather than a photo-voltaic type such as solar cells. When light impinges on the cell, its resistance decreases, thereby decreasing the peak voltage to which capacitor C2 can charge and its charging rate. Accordingly the first effect of light will be to *reduce* the load power. When sufficient light is received by the cell, the triac will be inhibited from triggering and power to the load will be turned off. An electrical heating or lighting system could conceivably be automatically turned off at dawn by exposing the cell to light from the rising sun. Various cadmium, sulphide, cadmium selenide, and selenium cells may be used. Thermistors may be similarly deployed to provide response to heat. It is also possible to place these sensors in the R1 portion of the circuit so that load power is *increased* in response to light or temperature.

WARNING

The RCA isolated-stud package thyristors should be handled with care. The ceramic portion of these thyristors contains beryllium oxide as a major ingredient. Do not crush, grind, or abrade these portions of the thyristors because the dust resulting from such action may be hazardous if inhaled.

SIMPLE SOLID-STATE AC RELAYS

Although power transistors and thyristors can be considered as solid-state versions of electromechanical relays or contactors, the analogy tends to be more academic than practical. This is because of the lack of electrical isolation between the input and output circuits of these devices. In actual relay applications it is comforting to know that there is considerable isolation between the control switch and the controlled circuit, which may involve voltages in the kilovolts range. Such isolation is readily brought about by means of an optocoupler, which usually consists of a light-emitting diode and a photosensitive device within a sealed package. When the optocoupler is used to turn on or off power transistors or thyristors, a good simulation of the erstwhile relay results. Indeed the electronic version can be endowed with superior operating features, such as zero-voltage "contact" closure and various logic functions. Even in its simplest forms the solid-state relay displays the advantage over physically actuated contacts in that no arcing occurs. If nothing else, this enhances reliability and longevity. Many different forms of the solid-state relay have been devised for both ac and dc control. Most of these utilize the optocoupler to achieve the desired input-output isolation.

Fig. 5-3 shows two simple solid-state ac relays. The circuit depicted in Fig. 5-3A simulates the normally open mechanical relay, whereas that of Fig. 5-3B is normally closed in the absence of input current. The use of the triac rather than an SCR for load switching is advantageous in that full-wave sinusoidal ac is delivered during the on state. (Where power considerations dictate the use of SCRs, a pair of such devices oppositely connected in parallel also yields full-wave control. However, the gate circuits then lack the simplicity made possible by the single gate of the triac.)

The 4N35 optocoupler is rated at 2,500-volts rms isolation between input and output. The solid-state lamp is an infrared emitting gallium-arsenide diode. The photosensitive device is a silicon phototransistor. The H11B2 optocoupler has an isolation rating of 660 volts rms and uses a silicon Darlington phototransistor. Both devices exhibit input-

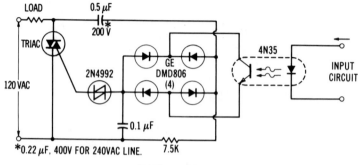

(A) Normally open.

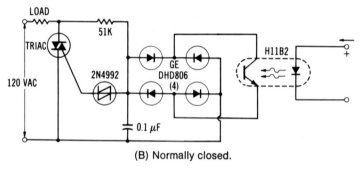

(B) Normally closed.

Fig. 5-3. Two examples of simple solid-state ac relays.

output isolation resistance in the vicinity of 100 gigaohms when the test is made with a 500-volt dc source. It is evident that safety considerations can be readily met with solid-state relays utilizing optocouplers. (Optocouplers are also available with isolation voltage ratings in the several thousands volts and higher.)

CONSTANT-BRIGHTNESS CONTROL

In photography, in various display projects, and in certain photochemical industrial processes, it is desirable not only to be able to control illumination brightness but to maintain a constant selected brightness level. The solid-state circuit shown in Fig. 5-4 can provide such brightness stabilization at any selected brightness level throughout the range of 100 percent to less than 10 percent of nominal lamp output. The basic circuit allows the choice of various 120-volt or 220-volt lamp sizes, because the circuit operates from *either* line voltage. The table for selecting

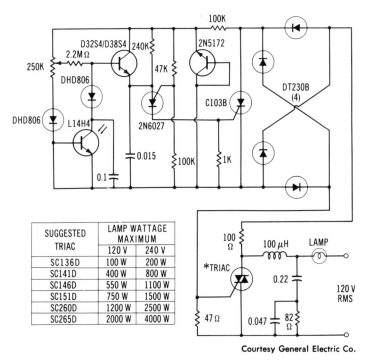

SUGGESTED TRIAC	LAMP WATTAGE MAXIMUM	
	120 V	240 V
SC136D	100 W	200 W
SC141D	400 W	800 W
SC146D	550 W	1100 W
SC151D	750 W	1500 W
SC260D	1200 W	2500 W
SC265D	2000 W	4000 W

Courtesy General Electric Co.

Fig. 5-4. Constant-brightness control circuit.

lamps and triacs is included with the schematic diagram. Due to the respectable amount of power that can be handled by this circuit, its range of applications can extend from indoor photography to the illumination of tennis courts and athletic fields.

In operation, light-sensing transistor L14H4 is exposed to a portion of the illumination emanating from the lamp. Such exposure constitutes optical completion of the feedback loop of the system. Therefore, this phototransistor should be physically situated in such a way that there is no interruption of or interference with the light impinging on it, such as might be imposed by fog or by dust clouds. The conductivity of the phototransistor, which varies with the amount of light it receives, in turn varies the conductivity of the D32S4/D38S4 transistor. This transistor serves as the resistive element of the RC relaxation oscillator configured about the 2N6027 programmable unijunction transistor. The output pulse from this relaxation oscillator triggers the C103B SCR, making it conductive at varying times in the ac cycle as governed by the

RC time constant of the PUT. (Recall that the D32S4/D38S4 transistor is the R element of the RC network.)

Before proceeding, two features of this circuit merit special comment. First, the C103B SCR, though triggered on by the relaxation oscillator, is commutated to its off state twice per ac cycle. This is because the SCR constitutes the "load" for the full-wave bridge rectifier circuit made up of the four DT230B diodes. Inasmuch as there is no filter capacitor, the anode-cathode voltage of the C103B goes to zero at a rate that is *twice* the power-line frequency. The second unique feature of this circuit pertains to the PUT relaxation oscillator. It too is forced to begin its oscillatory action anew each time the rectified dc from the bridge drops to zero. The 2N5172 transistor is actually used as a *zener diode*. It not only stabilizes the unfiltered dc applied to the PUT and to the phototransistor but also alters the waveshape in a favorable way. From what has been described, it can be seen that the actions of the PUT and the SCR are synchronized to the ac power line. Specifically the SCR is triggered on for a *certain portion* of the alternations of the ac power line. These on times are determined by the response of the phototransistor to the light received from the illuminating lamp.

Having traced circuit operation from the phototransistor to the C103B SCR, we will consider how control is extended to the triac and its lamp load. This final link is straightforward enough and is a technique frequently used in power electronics. Simply stated, if the output of a bridge rectifier is shorted, or is heavily loaded, the current must obviously be drawn from the ac line. In this circuit the path of the ac line to the bridge rectifier is through the 47-ohm and the 100-ohm resistors. Consequently, when the SCR is triggered into conduction, a heavier current flows through these resistances. The resultant increase in voltage dropped across the 47-ohm resistor triggers the triac into full-wave conduction. The lamp is thereby connected to the ac line. Furthermore, the triac is phase controlled because its gate responds to the phase-controlled SCR. In this manner the lamp current, and therefore the lamp brightness, is controlled in a closed-loop system.

Suppose that for some reason the light intensity begins to increase from a set brightness level. (This adjustment is provided by the 250K potentiometer.) The higher light intensity then impinging on the phototransistor *increases* its conductivity. In turn this diverts forward bias from the D32S4/D38S4 transistor, thereby *reducing* its conductivity. Thus, the PUT oscillator "sees" an *increased* RC time constant and takes a *longer* time to develop an output pulse. This means that the conductive states of both the SCR and the triac are delayed for a *longer* portion of the ac cycle and the rms value of the lamp current is *decreased*, as is its brightness. Conversely a reduction in brightness level is counteracted by the same sequence of oppositely directed events. In this manner any brightness level selected by the 250K potentiometer is regulated against line-voltage variations and changes in lamp characteristics.

In setting up such a control system, patient experimentation may be required to find the optimum distance between the phototransistor and the lamp. Once this distance has been determined, the brightness controller–regulator should be rigidly mounted so that the angular orientation of the phototransistor with respect to incident light cannot be inadvertently altered.

The 100-μH inductor and the RC network are included as an rfi filter to reduce injection of electrical noise from the triac into the ac power line.

SLAVE PHOTOFLASH UNIT

The heart of the photoflash circuit shown in Fig. 5-5 is a light-actuated SCR, or LASCR. This slave flash is superb for high-speed photography because the overall delay in the light burst is probably well under 100 microseconds. Several such units can be employed when photographic

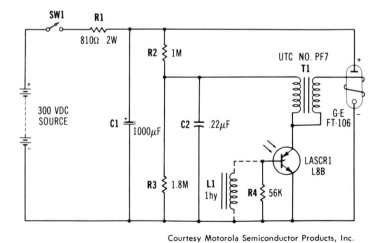

Courtesy Motorola Semiconductor Products, Inc.

Fig. 5-5.   Photoflash slave unit using an LASCR.

conditions are best served by extensive illumination. There is no need to be concerned with precise synchronization between the camera-actuated flash unit and the slave units because all slave units are triggered by the initial burst of light from the camera-actuated unit.

The LASCR actually has both electrical- and optical-gate provisions. The electrical-gate function is useful in controlling the optical *sensitivity* of the device. Assuming that the slave flash unit has been enabled by closing switch Sw1, energy-storage C1 charges to 300 volts within several seconds. This capacitor voltage is also impressed across the FT-106 photoflash tube. Approximately 200 volts are derived from capacitor C1 by the voltage divider consisting of R2 and R3. This voltage is applied to the LASCR circuitry. The slave flash unit will remain in this state until a burst of light impinges on the window of the LASCR. The LASCR will then be triggered in a fashion similar to ordinary SCRs when supplied with an electrical gate signal. The LASCR then "dumps" the charge accumulated in capacitor C2 across the primary winding of transformer T1. The secondary winding of the transformer develops a high-voltage pulse, which triggers the photoflash tube. Although turn-off of the LASCR might occur because of the momentary discharge of capacitor C1 by the ionized photoflash tube, turn-off is ensured by the resonant action between the primary winding of transformer T1 and capacitor C2. The first reverse-bias cycle of this shock-excited oscillation commutates the LASCR. Thus, the slave unit will be ready to fire again about as quickly as the master flash has recovered.

The gate circuit can be completed either through a resistance (R4), or through a 1-henry inductor. The latter arrangement is recommended. When the resistance is used, the LASCR may be vulnerable to triggering from ambient light. This can be remedied in some instances by reducing the value of resistance R4, but the sensitivity to flash actuation will also be reduced. Conversely, when the *inductor* is used, ordinary ambient light "sees" the LASCR with a nearly direct short in its gate circuit, and triggering sensitivity will be low. But with the fast and intense burst of flash illumination, the inductor will behave as a high impedance, thereby imparting high triggering sensitivity to the LASCR.

SIMPLE SCR BATTERY CHARGER

The battery charger circuit shown in Fig. 5-6 bears superficial resemblance to the half-wave SCR circuits commonly used for control of lamps

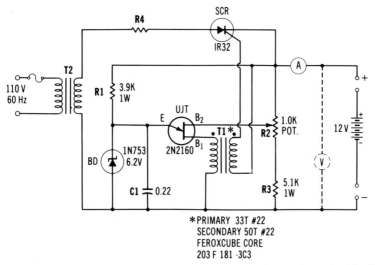

*PRIMARY 33T #22
SECONDARY 50T #22
FEROXCUBE CORE
203 F 181 -3C3

Courtesy International Rectifier

Fig. 5-6. Simple SCR battery charger with automatic shutoff.

and heaters. However, the mode of operation is quite different from that involved in the phase-shift control of 60-Hz power. The performance features are uniquely relevant to the needs of battery charging. Specifically this charger is immune to damage from short circuits and from inadvertently reversing the battery polarity. Best of all, the charger can be adjusted to shut down automatically when the battery attains its fully charged condition.

The circuit involves a unijunction transistor oscillator that delivers audio-frequency pulses to the SCR gate while the battery is undergoing charge. This results in substantially complete half cycles of 60-Hz current being delivered to the battery. (Inasmuch as the UJT oscillator frequency is much greater than that of the power line, there is always high probability that the triggering of the SCR will occur early in the excursion of the 60-Hz sine wave.) The SCR is self-commutated by the 60-Hz current; that is, conduction always ceases for 120th of a second despite the continual presence of the high-frequency trigger pulses.

As thus far described, the circuit operation is essentially that of a half-wave rectifier from the battery. Note, however, the presence of the 6.2-volt zener diode, BD. This diode cannot alter the operation of the oscillator as long as the peak-point voltage of the UJT is less than 6.2

volts. However, the peak-point voltage of a UJT increases with increasing interbase voltage. Accordingly potentiometer R2 can be adjusted so that the required peak-point voltage will be greater than 6.2 volts. In such a case the UJT will never be triggered, and oscillation will not occur. In turn, the SCR will not be triggered, and no current will be delivered to the battery from the 60-Hz power line.

In actual operation of the charger, control R2 is adjusted to enable oscillation to take place with charging current being delivered to a weak battery. As the battery accumulates charge, its terminal voltage rises. The interbase voltage of the UJT is thereby increased, and its required peak-point voltage increases above 6.2 volts. With proper adjustment of control R2, the oscillation of the UJT can be made to cease when the battery charges to about 14 volts. Thus, the charger will automatically shut down when the battery is fully charged, and unattended operation will not result in overcharging.

For many practical applications, transformer T2 can be a step-down transformer with a 15- to 25-volt secondary winding and a current capacity up to 15 amperes. Resistance R4 limits the maximum value of the charging current and is best determined experimentally. In most instances it will be a low-resistance, high-wattage, wirewound type. For initial evaluation of this charger, resistor R4 can be chosen as 0.5 ohm and 50 watts. An adjustable type will, of course, facilitate experimentation.

SOLID-STATE TEMPERATURE CONTROLLER

The electronic temperature controller or "thermostat" in Fig. 5-7 is capable of greater accuracy and better operating reliability than its mechanical counterpart. The triac is switched on or off by the dc voltage level appearing at its gate. The triac turns itself off each cycle because the ac load current is directly from the power line. Thus, there are no commutation problems in such an application. The load is either fully on or it is turned off. In an actual thermal system, the set-temperature adjustment is provided by the 50K potentiometer, and the duty cycle of load power will depend on the nature of the thermal equilibrium, which can be attained in the room or system. In this respect the operation simulates that of the bimetallic thermostatic switch.

The operation is crisp and positive because of an internal op amp within the IC, which is employed as a voltage comparator. The thermal

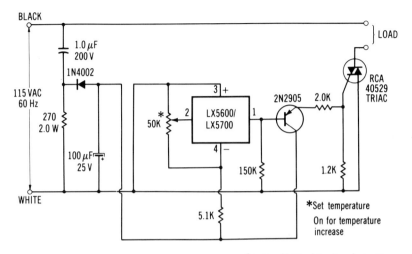

Courtesy National Semiconductor Corp.

Fig. 5-7. Temperature controller.

time constant of this IC is about 2 or 3 minutes in still air. In order to minimize errors from the effect of internal heating, a small heat sink can be clipped to the package. Best results are obtained with moving air rather than still air.

In the circuit diagram of Fig. 5-7 a 1.0-μF capacitor is used to drop the line voltage to the vicinity of 12 volts or so. The half-wave rectifier and filter are then made up of the 1N4002 diode and the 100-μF capacitor. Such a primitive power supply is adequate because there is a shunt voltage regulator within the IC. The designated triac is a sensitive-gate type and can control up to 2.5 amperes of load current.

The performance of this simple power control circuit is much superior to that readily attainable with discrete elements in the transducer function. There are more than two dozen transistors incorporated in the LX5600 IC, and the temperature-sensing technique is a sophisticated one. The basic temperature-sensing element uses the difference in emitter-base voltage of transistors operating at different current densities. The requisite close matching of this transistor pair is easily accomplished in a monolithic structure.

This circuit is amenable to the control of higher power loads. The experimenter can readily cascade driver transistors in Darlington circuits in order to provide the required gate drive for larger triacs.

This system does not deliver bursts of integral cycles of ac power to the load. Rather the load may be switched on at random times in the line-voltage waveform. Therefore, the possibility of noise generation exists. However, unlike phase-controlled systems, which generate continuous noise, a thermally actuated system such as this will only produce a short, and usually benign, burst of rfi or emi during the relatively infrequent turn-on time of the triac.

INTEGRAL-CYCLE PROPORTIONAL
TEMPERATURE CONTROLLER

At first the temperature controller shown in Fig. 5-8 might appear as just another version of thyristor-controlled load power with perhaps minor circuit modifications. The fact is, however, that the operational mode is significantly different from run-of-the-mill phase-control schemes or from simple on-off control schemes. This controller apportions whole numbers of *complete* ac cycles to the heater load by means of a zero-voltage switching technique. When a thyristor already has a gate signal applied as the ac cycle increases from its zero value, 180-degree conduction takes place. Because such triggering does not produce electrical noise, there is no need for an rfi filter in this controller. Another salient feature of this control method is that it constantly yields proportionate response. This manifests itself as more accurately maintained temperature control. The over-compensatory operation yielding "too-hot" and "too-cold" temperatures, inherent in mechanical and electrical off-on systems, tends to be of negligible magnitude in proportional integral-cycle control. (No thermal hysteresis need be *deliberately* incorporated in the system.)

Because of these features on integral-cycle control, proliferation of such circuitry can be expected in the future. Already a number of "zero-voltage switches" have made their appearance. These are sophisticated IC modules that perform the zero-switching function, and often only the thyristor and a few external components are needed to complete the circuit. The circuit of Fig. 5-8 is interesting in that the zero-switching circuit is made up of *discrete* components, which provides particularly clear insight into the *nature* of integral-cycle control.

The way in which the circuit of Fig. 5-8 operates is certainly not obvious from inspection. It is convenient for analysis initially to consider SCR Q3 as just an ordinary switch. Let switch Q3 be open and let the

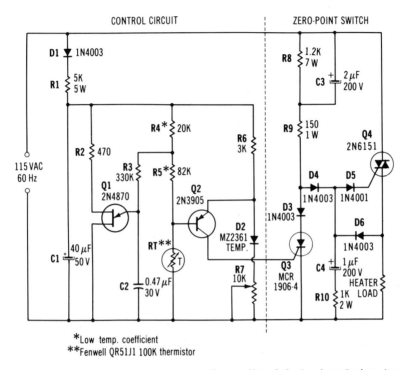

*Low temp. coefficient
**Fenwell QR51J1 100K thermistor

Courtesy Motorola Semiconductor Products, Inc.

Fig. 5-8. Proportional temperature controller using zero-switching technique.

circuit be connected to the power line. For the moment assume that the ac voltage cycle is just crossing zero in its excursion from negative to positive polarity. A path for the charging current of capacitor C3 will then be provided by resistance R9, diodes D4 and D5, and the gate of triac Q4. Therefore, triac Q4 will trigger close to the time that the 60-Hz voltage crosses the zero axis and will remain in conduction until self-commutated by the next zero-crossing one-half cycle later.

So far the circuit operation has been ordinary enough. However, the turn-on of the triac also initiates the charging of capacitor C4. This capacitor charges to the peak of the positive-going line voltage. As the line voltage declines from its peak positive value, diode D6 finds itself reverse biased by the charge stored on capacitor C4. Accordingly capacitor C4 begins its discharge cycle through diode D5 and the gate of the triac. This gate current persists while the ac voltage crosses zero as it traverses from positive to negative polarity. Therefore, the triac is also zero-voltage triggered for the negative half of the ac voltage wave. From the sequence of events described thus far, it can be seen that the triac will continue to fire at the half-cycle zero-crossing points of the applied ac voltage wave. Substantially sinusoidal load current and voltage will exist and, insofar as concerns load power, the situation is much the same as if the load were connected directly across the ac line.

If we consider the same initial circumstance but with switch Q3 closed, it should be plain that the triac would *not* be triggered as the positive excursion of the ac voltage wave develops. This is because gate current is diverted through diode D3 and switch Q3. It also should be apparent that capacitor C4 has no opportunity to charge during the positive half-cycle. Accordingly the triac does not fire during the negative half-cycle either. We should recognize a *unique* characteristic of this circuit—the closure of switch Q3 *inhibits* all triggering opportunities for the triac.

We next consider what happens if, as is likely to be the case, we are unable to close switch Q3 at the onset of the rising positive voltage wave. At this point in our discussion of circuit operation, no further convenience will be served by viewing Q3 as a simple knife-switch. Accordingly we will deal with it more realistically as an SCR capable of being turned on by a gate signal. If SCR Q3 is turned on by a gate signal at some random time during a *positive* half-cycle, the triac will remain conductive but will turn off at the end of the voltage half-cycle. However, if a gate signal is delivered to SCR Q3 at some random time during a *negative* half-cycle, SCR Q3 will not turn on at all because it is reverse biased. We arrive at the circuit logic whereby zero-voltage switching of the triac *must* occur whenever SCR Q3 is made to behave as an open switch. Relevantly the gate signal to SCR Q3 can be removed at any time. The triac will be triggered in all such instances but only when the ac voltage wave undergoes its negative-to-positive transition. To see why, consider the following reasoning.

Suppose that SCR Q3 has been turned on but that its gate signal is removed some time during the positive half-cycle of the voltage wave. By the nature of an SCR, Q3 will remain conductive for the entire positive half-cycle. This prevents the triac from turning on in the middle of such a positive half-cycle. Inasmuch as capacitor C4 does not have the opportunity to be charged, the triac will not turn on during the ensuing negative half-cycle either. Specifically the triac waits until the negative-to-positive zero-crossing to turn on.

Finally suppose that the SCR gate signal is removed during the negative half-cycle of the voltage wave. The SCR is already in its non-conductive state because of reverse anode-cathode voltage. Therefore, the SCR simply *remains* an open switch and the triac will not have its operation altered. The triac will ordinarily be on for the entire negative half-cycle and will then turn on for the forthcoming positive half-cycle. However, we can imagine yet another situation. Suppose that the ac power has been connected to the circuit during the negative portion of the voltage wave. In this instance both the SCR and the triac will be off. (Recall that the triac cannot be on during the negative voltage excursions if not on during the previous *positive* excursion. This is because capacitor C4 will not have had the opportunity to charge to gate-firing potential.)

The main point of these circuit responses is that power to the load can be controlled by varying the ratio of on time to off time of SCR Q3, because this will govern "dead time" and the number of ac cycles delivered to the load. If the load is a device with a large thermal time constant, such as a heater, the effect of such control will be similar to that which would result from a variation of the applied voltage or current.

Having investigated the all-important technique of *zero-voltage* switching, we are in a favorable position to use it for controlling temperature. Referring again to Fig. 5-8, a simple half-wave dc power supply is used for the control circuit. This dc power supply is made up of rectifier diode D1, voltage-dropping resistance R1, and filter capacitor C1. For the moment, circuit operation will be analyzed exclusive of the relaxation oscillator, UJT Q1. Thermistor RT, resistances R4, R5, R6, R7, and diode D2 can be seen to form the arms of a Wheatstone bridge. Transistor Q2 may be considered the null detector for this bridge network. Variable resistance R7 is set so that bridge balance prevails at a desired temperature. As room or ambient temperature increases, the resistance of thermistor RT decreases. Ultimately transistor Q2 will be turned on and will provide gate drive to SCR Q3. This, as explained, turns off the triac and therefore the heater power. The reverse sequence of events occurs when the room temperature drops.

As described, the system is workable in that the heater will be turned on and off in such a way that an average temperature will be maintained as set by the adjustment of R7. Random bursts of integral cycles will be delivered to the heater, and no rfi will be produced.

However, the overshoot and undershoot of temperature can be at least several degrees, depending on thermal conditions of the environment. Much better control resolution can be attained by *modulating* the "error" signal. This is the purpose of the UJT oscillator, Q1. It adds a sawtooth voltage to the bridge network. The period of this superimposed sawtooth signal is designed to equal 12 cycles of the power-line frequency. The sawtooth wave modulates the bridge voltage so that the net bridge voltage will be above null for a fraction of the 12-cycle group and below null for the remaining fraction of the 12-cycle group. The actual proportioning of on and off cycles within the 12-cycle groups depends on how far the temperature has strayed from nominal value. Such temperature deviation is "reported" by variations in the resistance of the thermistor. The operational tendency of such control action is for the duty cycle of the on-off periods to stabilize. Temperature is then maintained over a narrow range.

Additional zero-voltage switching circuits are shown in Fig. 5-9. These accomplish the zero-voltage switching process with the use of specialized ICs. The proportional temperature controller of Fig. 5-9A functions similarly to the discrete-component circuit in Fig. 5-8. Here again, the null point of the error signal is modulated by a sawtooth waveform. However, the relaxation oscillator is configured around a programmable unijunction transistor (PUT) rather than a conventional UJT device. Because of the high-wattage heater controlled by this system, the sawtooth period is relatively slow. The oven heat controller of Fig. 5-9B employs an RCA zero-voltage switch IC known by the acronym *ZVS*. Three of these zero-voltage switches are the CA3058, the CA3059, and the CA3079. The circuit shown in Fig. 5-9B does not impart modulation about the set point. Because of the closely lumped thermal "circuit" in an oven, simple zero-switching control suffices. A general-purpose zero-voltage-crossing power switch is shown in Fig. 5-9C. The operation of the MFC8070 IC is such that the *opening* of switch SW applies power to the load at the first zero-voltage crossing. If desired, an opto-isolator can be substituted for the mechanical switch indicated. In this way safe electrical isolation will be provided between the control switch and the power circuit.

Another worthwhile feature of power control by zero-voltage switching is that the thyristor is subjected to the minimum possible rate of current change when applying power to a steady-state load. That is,

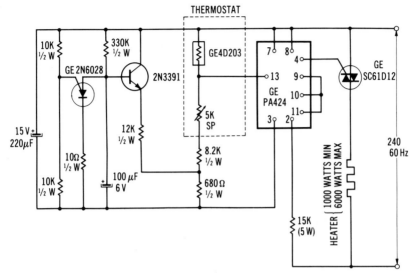

Courtesy General Electric Co.

(A) High-power proportionate temperature controller using the
General Electric PA424 IC.

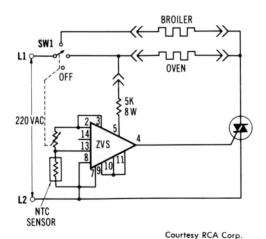

Courtesy RCA Corp.

(B) Oven heat controller using an RCA zero-voltage switch IC.

Fig. 5-9.   Power control circuits using zero-voltage switching ICs.

Fig. 5-9. *continued*

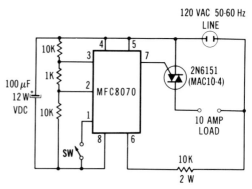

Courtesy Motorola Semiconductor Products, Inc.
(C) On-off switch using the Motorola MFC8070 IC.

di/dt is as safe as it can be. Keep in mind, however, that the systems and circuits thus far described work best with resistive loads.

FLASHERS, CHASERS, AND COUNTERS

Various patterns of blinking lights are widely used for advertising, displays, traffic lights, hazard alerts, aircraft and auto signaling, etc. Before the availability of reliable solid-state power devices, these functions were typically accomplished with motor-driven cams and heavy silver contacts. At their best, such mechanical systems are objectionable because of arcing and because of maintenance due to the physical deterioration of the contacts.

Of the three commonly used sequential patterns for switching lamps, the *flasher* represents the simplest technique. Here one or two lights alternately flash on and off at a predetermined rate. In *counters,* several lamps are arranged in a linear or other geometric array and only one lamp is kept on at a time. However, there is a continuous sequential transfer of light from one lamp to another. The process is suggestive of a "bucket brigade," although to the observer the "passed" entity appears to be light rather than buckets. The *chaser* also involves progressive movement of light. In this operational mode the first and all following lamps *remain* on. Then, after all lamps are on, they all are simultaneously turned off and the entire process recycles, commencing with the first lamp. To the observer an ever-lengthening "arrow" appears to be chasing the remaining dark segment of the display.

Flashers

The flasher circuit shown in Fig. 5-10 is representative of modern trends in power control. By utilizing dedicated IC modules, both circuit simplicity and performance are enhanced. The RCA ZVS ICs are specifically intended to accomplish zero-voltage switching with thyristors. Zero-voltage switching greatly reduces rfi and emi. Additionally it enables the triacs to operate under relatively moderate conditions—gate signals are provided at the zero-voltage crossing of the ac cycle and the triac is spared high rates of current change. This is particularly significant in switching incandescent lamps where the cold resistance of the filament is much lower than the hot resistance. No dc supply is required for these ICs inasmuch as they contain their own internal dc system.

In some respects the zero-voltage switch may be thought of as a specialized op amp. For example, ZVS1 in Fig. 5-10 is connected to an external RC network, which causes it to function as an astable multivibrator. Note the connection between terminal 6 on ZVS1 and terminal 1 on ZVS2. Terminal 6 on these ICs is simply a logic-level output, whereas terminal 1 provides an "inhibit" function. Referring to the flasher circuit, the voltage at terminal 6 of IC ZVS1 is high when lamp 1 is on and ZVS2 is thereby inhibited from operation during this interval. Thus, the self-oscillating IC, ZVS1, is the *master* and ZVS2 is the *slave*. When lamp 1 is turned off at the end of one-half period of the multivibrator cycle, ZVS2 is no longer inhibited and will turn on lamp 2 at the first available zero crossing of the 60-Hz line voltage. Thus, lamps 1 and 2 alternate their on and off times.

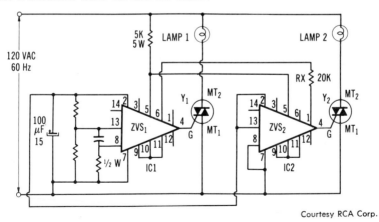

Fig. 5-10. Solid-state flasher with zero-voltage switching and triac lamp control.

An initial "feel" for the flashing rate may be extrapolated from the following information. If the two resistors connected in series between terminals 2 and 7 of ZVS1 are each 100K and the series RC network associated with terminals 7, 8, and 13 includes a 10K resistor and a 10-μF capacitor, the *on* time of each lamp will be in the neighborhood of 3.5 seconds.

The triacs are the RCA type T2800B, which permits the use of 300-watt lamps.

Chasers

The chaser circuit of Fig. 5-11 is intended for advertising displays where the illusion of a "moving" sign is desired. This particular arrangement has the feature that the rest time during recycling of the lighting sequences is twice the duration of the intervals between the turn on of subsequent lamps as the light pattern develops. It is felt that such a rhythm is more appealing to the eye. However, this feature can be readily dispensed with if so desired.

When the system is connected to the ac power line, all of the SCRs will be initially turned off. The unijunction transistor oscillator configured about Q1 will then begin to produce the first trigger pulse. This will require an amount of time determined by the charging rate of timing capacitor C2. One might suppose that all of the SCRs will be turned on by this first trigger pulse because their gates are connected through diodes to the trigger oscillator. This, however, is not the case.

Actually turn-on will be *inhibited* in all SCRs except SCR1. The reason for this is that a dc charge trapped in coupling capacitors C8, C9, C10, and C11 reverse-bias the gate diodes, CR5 through CR8. Indeed it is the manipulation of these trapped charges that leads to the desired sequencing of the SCRs and therefore the triac lamp circuit. These trigger-inhibiting charges become trapped in the coupling capacitors under two conditions:

1. When the chaser circuit is first connected to the ac power line
2. When the last SCR (SCR5) is turned off

The trapped charges in the coupling capacitors are *depleted* when a preceding SCR is turned on. Thus, each SCR involved in the sequencing pattern is "cocked" ahead of time so that its gate diode is able to conduct

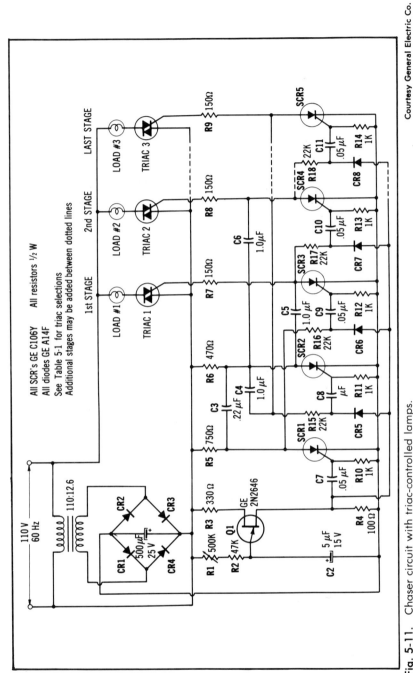

Fig. 5-11. Chaser circuit with triac-controlled lamps.

Courtesy General Electric Co.

267

the trigger pulse when it comes along. The lamp-controlling triacs are in turn triggered by SCRs 3, 4, and 5. Triac conduction endures as long as its controlling SCR remains on. When such an SCR is turned off, its triac automatically turns off because of the commutating action of the 60-Hz line voltage.

The problem remains of turning off the SCRs. Inasmuch as these thyristors are operated from filtered dc, they are not naturally commutating and they would tend to stay in their turned-on state indefinitely. It can be seen that SCRs 3, 4, and 5 have their anodes connected to the anode of SCR2 through a capacitor. Such interanode capacitors provide commutation to the already turned-on SCRs when SCR2 is turned on. (A turned-on SCR develops a negative-going pulse, which is transferred through the interanode capacitor and momentarily deprives the already turned-on SCR of holding current, thereby turning it off.)

With this discussion as prelude, let us trace the circuit operation through a full cycle of lamp sequencing. When the flasher is connected to the ac power line, all thyristors are initially in their off state. When oscillator stage Q1 develops its first trigger pulse, SCR1 will be turned on. No trigger will be delivered to the other SCRs because their gate diodes (CR5 through CR8) will be reverse biased, as explained. However, when oscillator stage Q1 produces its *second* trigger pulse, SCR3 will be turned on. Therefore, triac 1 together with lamp load 1 will turn on. The reason for this is that the previous turn on of SCR1 depleted the trapped charge in coupling capacitor C9, therby disposing of the reverse bias on gate diode CR6. The turn-on of SCR3 depletes the stored charge in coupling capacitor C10, thereby "cocking" SCR4 for the third trigger pulse. When the *third* trigger occurs, SCR4 is turned on, as is its triac 2 and lamp load 2. The *fourth* trigger pulse finds SCR5 receptive to a gate signal and that SCR is then turned on, as is triac 3 and lamp load 3.

Turn-on of lamp 3 produces a light pattern with all three lamps glowing. The last described event, the turn-on of SCR5, depletes the reverse voltage previously stored in coupling capacitor C8, enabling gate diode CR5 to be receptive to the next trigger pulse. Accordingly the *fifth* trigger pulse turns on SCR2, which then turns off SCR3, SCR4, and SCR5 via commutating capacitors C5, C6, and C4 respectively. At this point in the sequence, *all* lamps are off, and coupling capacitors C8 through C11 have again acquired trapped charges that inhibit their gate diodes (CR5 through CR8). The flasher is ready to undergo another cycle of operation.

The commencement of a new cycle of operation requires *two* trigger pulses. This accounts for the *prolonged* rest period between successive cycles of operation.

Modifications can be readily made to the mode of operation just described. For example, an additional triac and lamp load can be associated with SCR1 by connecting the triac gate through a 150-ohm resistor to the anode of SCR1. A four-lamp chaser will then be obtained with the rest period between cycles equal to the interval between turn ons. Alternately SCR1 and its associated components can be deleted. The system will then be a three-lamp chaser with equal intervals between lamp turn-ons and between pattern cycles.

Table 5-1 shows some triac and lamp combinations that may be used. The circuit in Fig. 5-11 is quite flexible and other than the indicated components may be employed, although it is suggested that the substituted semiconductor devices have characteristics similar or superior to those indicated. Also any number of stages and lamps can be inserted between SCR4 and SCR5 as long as the dc supply does not become overloaded.

TABLE 5-1. Recommended Triacs for Incandescent Lamp Loads

Maximum Wattage for 120 V Operation	Package		
	Plastic	Stud Mounted	Press Fit
15 Watts*	C103		
150 Watts*	C106		
700 Watts	C122	C20	C22
2500 Watts		C30	C32

*Must be used with a ballast resistor to ensure adequate surge protection.

Counters

A somewhat similar circuit to that of the chaser is shown in Fig. 5-12. However, a few subtle differences in the circuit configuration makes this system operate as a *counter* rather than a chaser. The lamps light up sequentially, but whenever a new lamp comes on, its predecessor turns off. The visual effect is one of a traveling light source rather than a lengthening bar of light. If, for example, a number of lamps were arranged in a circle, one would observe a rotating light making uniformly timed angular jumps along the arc of the circle. (If such lamps were

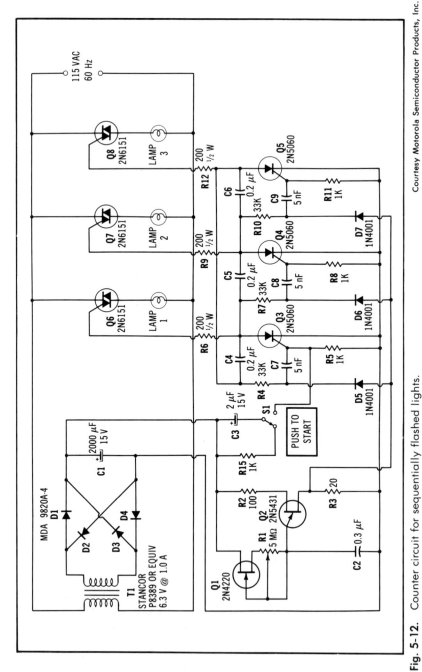

Fig. 5-12. Counter circuit for sequentially flashed lights.

Courtesy Motorola Semiconductor Products, Inc.

driven by a chaser circuit, the visual effect would be as if a jerky hand with a "light pen" repetitively drew and extinguished a circle of light.)

The four component sections of this circuit are readily identifiable. They are the *dc power supply,* the *display section,* the *trigger oscillator,* and the *shift register,* or *ring counter.* The dc power supply consists of power transformer T1, the bridge rectifier involving diodes D1 through D4, and filter capacitor C1. The display section contains the indicated triacs and their associated lamps. The trigger oscillator includes junction FET Q1 and UJT Q2. The ring counter involves SCRs Q3, Q4, and Q5 and associated circuitry.

Before describing the operation of the system, mention will be made of two unusual aspects of the circuitry. The UJT oscillator utilizes field-effect transistor Q1 in place of the more commonly encountered resistor to charge timing capacitor C2. Actually the Q1 circuit functions as a constant-current source, with the amount of current "metered" out to C2 determined by the adjustment of R1. (Although R1 may appear to be in the same circuit position as the conventional charging resistance, the fact that the current must flow through the FET makes a considerable difference.) Specifically this technique enables a greater range in the trigger-pulse rate than is readily obtained with simple resistance control. The approximate range of the trigger-pulse rate for this oscillator is from one every 0.1 second to one every 8 seconds.

The second unusual feature of this circuit is the way that the SCR anodes receive their positive dc voltage. Before any of the triacs can fire, it is necessary that the SCRs have an operating dc voltage applied. This dc voltage is obtained through the gate circuit of the triacs but in the opposite direction used to trigger the triacs. Inasmuch as a triac gate can be triggered with *either* polarity, this scheme may appear perplexing. However, the SCRs in their off states do not allow sufficient gate current to fire the triacs. Only an SCR that has been turned on can fire a triac and thereby turn on its associated lamp.

When the counter is first connected to the ac power line, the oscillator will commence operation, but there will be no visual evidence of this because all the lamps will remain off. This is because the trigger pulses will not be able to pass through the SCR gate diodes, D5, D6, and D7. These diodes will be reverse biased due to their connections to the gates of the triacs. (It was previously explained how a small current passed from the positive side of the power supply through the triac gate circuits.) The SCRs are *otherwise* ready to be triggered because they have nearly 9 volts across their anode-cathode terminals.

To place the circuit in operation, the push-to-start switch is momentarily depressed. This provides sufficient gate current from the discharge of capacitor C3 to trigger SCR Q3. As SCR Q3 turns on, its current demand causes the triggering of triac Q6, and lamp 1 is turned on. At the same time that SCR Q3 fires its triac, it also *removes* the reverse bias from gate diode D6. Thus, SCR Q4 becomes "cocked," and the *next* trigger pulse then turns on Q4. Because of commutating capacitor C5, SCR Q3 is turned off when SCR Q4 is turned on. Then SCR Q4 simultaneously fires triac Q7, and lamp 2 comes on as lamp 1 is extinguished. At the same time the reverse bias is removed from gate diode D7. This circuit action continues with SCR Q5 and triac Q8 being turned on at the next trigger pulse. The turn-on of triac Q8 is accompanied by the lighting of lamp 3 together with turn-off of lamp 2. Thus, the lighting sequence has been lamp 1, lamp 2, lamp 3. Inasmuch as the circuitry associated with SCR Q5 connects back to that of SCR Q3, the SCR control circuit is "circular." Therefore, the lamp sequence will be continuous and will indefinitely generate the light pattern, 1, 2, 3, 1, 2, 3, 1, etc.

Although only three stages are shown, additional SCRs, triacs, lamps, and associated components can readily be added to the indicated circuitry. Lamps up to about 500 watts can be controlled by the triac types shown. This power level is somewhat below the rated capability of these triacs, but allowance must be made for the high initial currents required when cold incandescent lamps are turned on.

FLUORESCENT LAMP DIMMER

The control of light output from fluorescent lamps is desirable for many applications. The fluorescent lamp dimmer about to be described does not use the most sophisticated electronics available. However, the circuit is basically satisfactory and is readily amenable to modifications once the underlying principles are understood.

The block diagram illustrating the basic concept of a fluorescent lamp dimmer is shown in Fig. 5-13A. The unique characteristics of the fluorescent lamp require that its light output be controlled by *current* variation rather than by changing the lamp voltage. This is because gaseous discharge lamps establish a nearly constant ionization potential across their terminals. It would not be practical to attempt to forcibly change this inherent feature. Fortunately, current variation will occur naturally if the traditional series ballast impedance used with fluorescent lamps is retained. This is because the ballast, a high-impedance inductor, absorbs across its terminals the *difference* between the applied peak voltage and the voltage needed to maintain ionization of the lamp. It is also fortunate that SCR phase control is essentially a duty-cycle modulation technique. (It is not perfect in this respect because the peak amplitude of the "chopped" line-voltage wave is not constant. However, the lamp will remain illuminated as long as the peak amplitude is sufficient to maintain ionization. In practice a wide range of light-intensity control is attainable.)

It is phase control of line voltage in conjunction with the ballast impedance that enables the average value of lamp *current* to be varied. It also readily supplies the *higher* voltage demanded by the lamp for starting. However, there is yet *another* requirement for such a control circuit. If half-wave phase control were used, the dc component in the controlled pulses would saturate the iron core of the ballast transformer. This would decrease its impedance and would interfere with the objec-

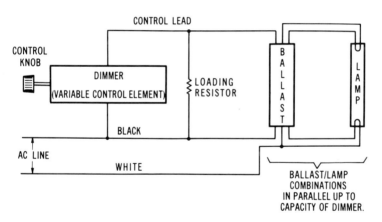

(A) Block diagram.

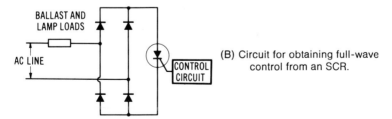

(B) Circuit for obtaining full-wave control from an SCR.

Fig. 5-13. Block diagram and circuit for fluorescent lamp dimmer.

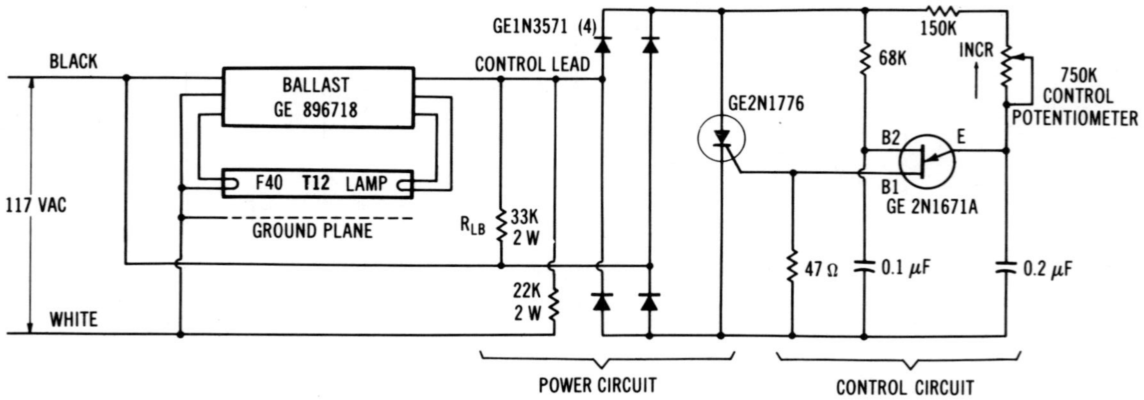

(C) Complete dimmer circuit.

Fig. 5-13. *continued*

Courtesy General Electric Co.

tive of controlling the lamp current. Therefore, what is needed is full-wave phase control, which, because of its waveform symmetry, has no dc component. Fig. 5-13B shows how full-wave control is obtained from a single SCR when it is used in conjunction with a bridge rectifier. (Of course, the experimenter may wish to use a triac instead of an SCR.)

The control circuit shown in Fig. 5-13C is essentially that of a unijunction transistor used as trigger control for an SCR. As in many other UJT–SCR applications, phase control of the SCR is achieved as the result of the time required to charge the emitter capacitor to the firing voltage of the UJT. In this circuit the 0.2-μF emitter capacitor is charged through the 150K resistance and the 750K control potentiometer.

The 33K loading resistance, R_{LB}, is needed to promote control stability. Its actual value should probably be investigated empirically. Unless R_{LB} is low enough to make the load appear sufficiently resistive, the SCR "sees" an inductive load and may encounter commutation problems.

The number of such lamp-ballast combinations that can be controlled by this system depends, of course, on the current capability of the SCR and the bridge rectifier. Also it may be found that better results are obtained from lamps that have been "seasoned" somewhat, rather than from brand-new lamps. Lamps in depressed-temperature environments, or those exposed to chilling drafts, may display unreliable ionization characteristics. They may refuse to start or they may not provide a satisfactory range of light control. Special low-temperature lamps should be used in such instances.

A HIGH-FREQUENCY FLUORESCENT LIGHTING SYSTEM

The light-producing efficiency of fluorescent lamps is better at frequencies sufficiently high that the gas does not have time to deionize between the ac voltage alternations. More important, it is contended that such light constitutes higher quality illumination. The threshold of flicker perception from lamps operating from the 60-Hz line may be close enough to the response of the eye to cause the headaches and discomfort some people apparently experience. Also high-frequency fluorescent lighting may prove better for certain photographic purposes.

Fig. 5-14 shows an 8-kHz inverter with sufficient power capability to handle six 80-watt instant-start fluorescent lamps. This inverter uses high-speed SCRs and is intended to be driven from a variable-duty-cycle trigger source. The trigger source is shown in Fig. 5-15. Such pulse-width control is an ideal way of controlling the average current and

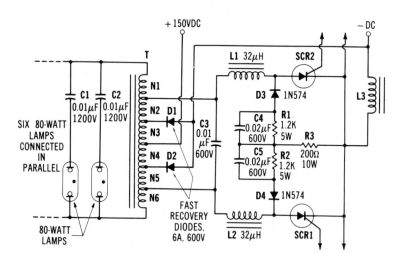

T – CORE, 8 PIECES OF INDIANA GENERAL NO. CF-602 MATERIAL 05, OR EQUIVALENT.
CROSS SECTION, 8 cm²
N1, N6 - 30 TURNS OF NO.18 MAGNET WIRE
N2, N5 - 13 TURNS OF NO.18 MAGNET WIRE, 2 STRANDS
N3, N4 - 52 TURNS OF NO.28 MAGNET WIRE, 2 STRANDS
L3 – 131 TURNS OF NO. 15 MAGNET WIRE ON ARNOLD
ENGINEERING CORE NO. A4-04117, OR EQUVIALENT.

Courtesy RCA Corp.

(A) Inverter circuit.

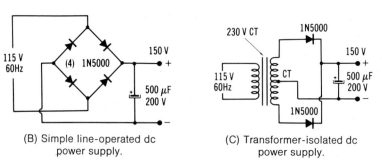

(B) Simple line-operated dc
power supply.

(C) Transformer-isolated dc
power supply.

Fig. 5-14.   High-frequency inverter for fluorescent lamps and two optional power supplies.

therefore the light output of fluorescent lamps. The inverter and its trigger source are intended for operation from a nominal 150-volt dc source with about 4-ampere capability. Such a dc supply voltage is readily obtained by rectification of the power-line voltage as shown in Fig. 5-14B. However, this simple scheme introduces the possibility of grounding problems unless precautions are taken to "float" the lamp, inverter, and trigger circuits. A safer supply employs an isolation transformer, as shown in Fig. 5-14C. Such a transformer should have a 600 VA rating in order to provide the current demand.

The inverter is the parallel type, and its operation is derived from the use of the high-speed SCRs together with a reliable commutation technique. The commutating capacitor is C3, but L1, L2, C4, C5, R1, R2, D3, and D4 are also involved in bringing about precise turn-off of one SCR when the alternate SCR turns on. Diodes D1 and D2 prevent voltage overshoots during switching transitions. Inductor L3 serves as a ballast to limit current peaks,which tend to occur during switching. These circuit features are more or less common to SCR inverters, but in this application they have been optimized to ensure reliable commutation with 8-kHz variable-duty-cycle operation.

An interesting aspect of this system is the use of capacitors, rather than inductors as ballasts for limiting lamp current. These capacitors have another function as well. They remove the dc component from the waveform delivered to the lamps. This is why control of lamp current and therefore light intensity is feasible. As the duty cycle of the inverter output wave departs more and more from 50 percent (by adjusting R5 or R11 in the trigger circuit in Fig. 5-15), an increasingly greater amount of energy is invested in the dc component. (The symmetrical square wave with 50 percent duty cycle has no dc component to begin with.) Because ballast capacitors C1 and C2 displace the zero axis in such a way as to make the dc component zero, the average current in the lamps varies with duty cycle. If it were not for this action, that is, if resistors rather than capacitors were used as the ballast elements, duty-cycle variation would have little effect on lamp operation because the net energy in the waveform delivered to the lamps by autotransformer T would remain constant. It is also evident that the use of inductive ballasts would not be appropriate in this circuit. (Inductors could be used, however, if the *control* of light intensity is not required.)

The trigger circuit of Fig. 5-15 is designed around an 8-kHz multivibrator in which transistors Q2 and Q3 are the active devices. Transistors Q1 and Q4, together with the neon lamp and associated components,

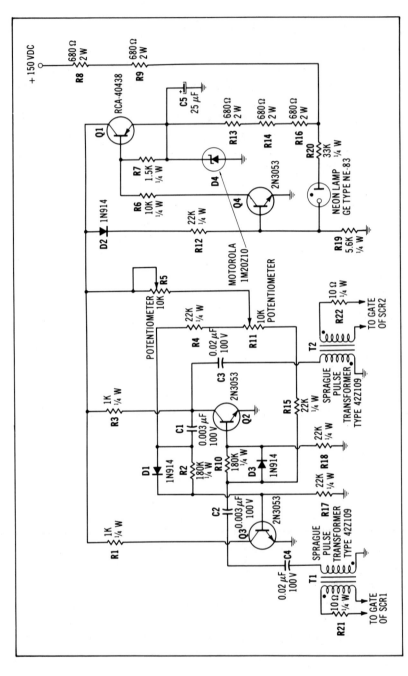

Fig. 5-15. Trigger circuit for high-frequency fluorescent lamp inverter.

Courtesy RCA Corp.

278

comprise a voltage-regulated power supply for the multivibrator. Transistor Q4 is the error amplifier–comparator, transistor Q1 is the series-pass transistor, and the neon lamp is the voltage reference. Zener diode D4, filter capacitor C5, and resistors R8, R9, R13, R14, and R16 form a 20-volt partially regulated dc supply. Therefore, the voltage and power rating of transistor Q1 can be reasonably low.

The salient feature of the multivibrator circuit is its duty-cycle control, which is provided by potentiometer R11. Potentiometer R5 is used to adjust the duty-cycle *range*. Once this range has been determined, R5 can be replaced with a fixed resistor, if desired. The two multivibrator outputs are obtained from the secondary windings of pulse transformers T1 and T2.

COLOR LIGHT ORGAN CIRCUIT

Colored lights flickering in response to music can add another dimension to perception. Sophisticated color organs have been developed in which there are certain unique correlations between audio frequency, loudness, and the color and intensity of responsive light sources. It is even conceivable that the right kind of luminous display could create the feeling of the music with no sound impinging on the ears. However, the novelty aspect of dancing lights synchronized to music is rewarding in itself. Power electronics is very much involved because the light displays, whether they be filamentary, gaseous, or solid-state lamps, must be actuated and controlled by power devices. (Thus far, LEDs have not been used much for this purpose because the eye demands strong sources of light.)

The circuit shown in Fig. 5-16 is a basic building block for experimentation with simultaneous aural and visual effects of sound. To get the feel of this phenomenon, it is suggested that initial investigation be confined to the simplest situation in which a single light source is controlled by monophonic sound. Later, when one has developed insights into relationships between sound and light, more complex multicolored systems can be worked out for stereo reproduction.

The ideas incorporated in the circuit of Fig. 5-16 are straightforward and are representative of much-used techniques for lamp dimmers, motor controls, etc. A selected portion of the audio spectrum enters the circuit through the input filter network. Actually, only positive portions of audio signal enter the circuit due to the diode inserted between

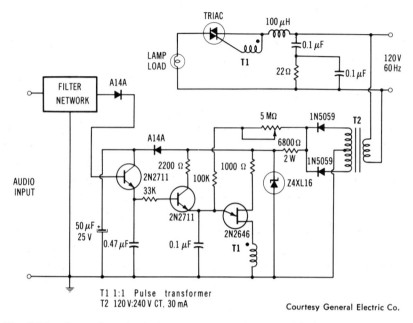

Fig. 5-16.   Circuit for color light organ or sound-activated lights.

the filter and the emitter-follower input stage. (Although 2N2711 transistors are shown in this schematic, many npn silicon audio transistors can be used.) The unidirectional pulses from the emitter follower are then applied to the base of the second transistor, which operates as an "electronic resistance." That is, its effective collector-emitter resistance is modulated by the amplitude of the applied base signal. Although it can be said that many transistors operate in this manner, it is particularly significant in this circuit because it is desired to vary the rate of charge of a capacitor in a relaxation oscillator. In other words an RC time constant is varied in accordance with audio information.

The time constant of this RC network determines how soon the triac will be triggered in the 60-Hz cycle. Thus, the audio signal is able to phase-control the triac and thereby vary the effective value of the lamp current. It should not be supposed that this occurs for each cycle of the audio signal. The 0.1-μF capacitor in the oscillator circuit cannot change its voltage instantly even though the time constant is changing. Rather the average effect of the music over a fraction of a second or more causes the light variations. This effect is also accomplished by the thermal lag of the lamp filament. (In systems using gaseous lamps, the effect is noticeably different because of the relatively fast response of these lamps.)

The usual approach to a color-organ display is to have three or more different colored lamps. Each of these lamps is driven by a separate circuit such as that of Fig. 5-16. The filter response for each circuit is different, however. For example, a red lamp might be driven from a circuit employing a 40- to 400-Hz bandpass filter, a green lamp driven from a second circuit using a 400- to 4,000-Hz filter, and a blue lamp driven from a circuit with a 4,000- to 20,000-Hz filter. (The low-frequency filter can alternatively be a low-pass type with a 400-Hz cutoff, and the high-frequency filter can be a 4,000-Hz high-pass type.) Because filters that are both economical and practical may not have steep attenuation slopes, the design is often made for much narrower bandwidths than those just suggested. This partially takes into account the fact the high-amplitude audio signals will ride through the stopband regions of simple filters. Thus, with narrower bandwidths, there will be *less* simultaneous illumination of the lamps.

A few words are in order with regard to the input filter. From the viewpoint of the experimenter, LC filters impose practical difficulties in altering response characteristics. Active electronic filters alleviate this problem. With recently available ICs containing four operational amplifiers on a single monolithic chip, the circuit complexity of active filters has been greatly reduced. Thus, it turns out that the LM124 quad op amp is ideal for electronic filters. The circuit for such a filter is shown in Fig. 5-17. The LM124 connection diagram is indicated in Fig. 5-18.

With the component values indicated in Fig. 5-17, a bandpass response will be obtained with center frequency at 1,000 Hz and a bandwidth of about 20 Hz. Once the filter is operative in this manner, it is easy to modify its behavior. In order to control bandwidth, a 500K variable resistance should be connected from point A to ground. Center frequency control can be attained by replacing R5 and R6 with 2-meg-ohm controls. Preferably these controls should be ganged on a single shaft. Additionally the center frequency can be shifted by changing the values of capacitors C1 and C2. However, these capacitors should be maintained equal in value.

It will probably be desirable to associate either an input or output potentiometer with the active filter in order to provide amplitude control. This is because the overall gain of the circuit changes with the

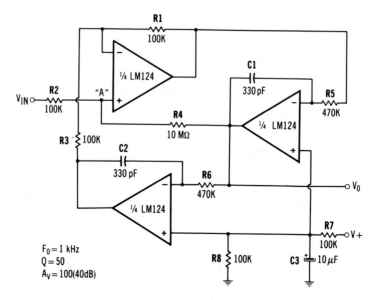

Courtesy National Semiconductor Corp.

Fig. 5-17. Basic active filter for use with color organ circuit in Fig. 5-16.

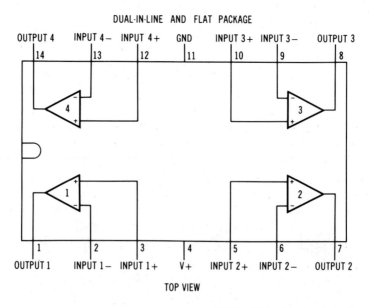

Courtesy National Semiconductor Corp.

Fig. 5-18. Connection diagram for the LM124 quad operational amplifier.

adjustment of bandwidth. (Much higher gains are developed at narrow bandwidths than for wide bandwidths.) An alternative approach would be to use the fourth op amp as a variable-gain output stage.

When experimenting with the color-organ circuit, always ascertain that the filter is performing as intended before evaluating the performance of the entire circuit. Also bear in mind that electronic filters are vulnerable to overload.

CONTROL CIRCUIT FOR SHIFTING BETWEEN TWO LIGHT SOURCES

The circuit of Fig. 5-19 provides the useful feature that light output can be smoothly shifted from one lamp load to another. Known also as a *tandem dimmer,* its main application is for slide projectors. However, the circuit can be readily used for stage effects and for advertising displays. It comprises two phase-controlled triac circuits, which share the phase-adjust control, R3. The adjustment of control R3 is such that the conduction angle of one triac is advanced while the conduction angle of the other triac is retarded. Thus, lamp current can be continuously transferred from one lamp load to the other as control R3 is varied.

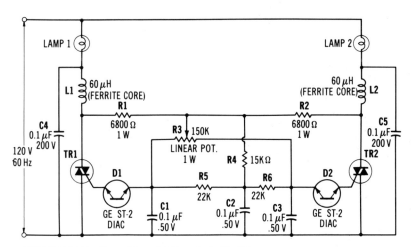

NOTE: Total light level (sum of lamps 1 & 2) constant within 15%.

Courtesy General Electric Co.

Fig. 5-19. Control circuit for smoothly shifting from one source to another.

This lamp control scheme, like other thyristor phase-control circuits, must be able to survive the heavy current inrush caused by the low resistance of cold lamp filaments. For this reason it is not sufficient to select triacs merely on the basis of the hot filament-current demand. Moreover, the available short-circuit current from the ac source enters into the picture. A different situation is presented by ordinary residential installations than by industrial or commercial power systems. Table 5-2

TABLE 5-2.   Triac Selection Guide for Phase-Controlled Lamps

Device	House Wiring	Industrial/Commercial Wiring
SC35/36	360 W	180 W
SC40/41	600 W	250 W
SC141	600 W	480 W
SC45/46	1000 W	450 W
SC146	1000 W	600 W
SC50/51	1200 W	600 W
SC60	2000 W	1300 W

Courtesy General Electric Co.

provides useful guidance in matching the triacs to various lamp loads. Note that the higher resistance of house wiring enables triacs with lower power ratings to be employed.

It is appropriate to mention another aspect of thyristor operation in this and in other lamp-control circuits. It may appear that fusing is unnecessary because of the low probability of a shorted load. However, the ordinary advent of a burned-out lamp can be nearly as destructive to the thyristors as a "crowbar" short circuit. This is because the parting filament may be momentarily followed by an arc. Although such arcing will endure only for the remainder of the half-cycle of energization, its inception can destroy thyristors. Table 5-3 is a guide for selecting fuses that will protect certain thyristors against such destruction. The basic idea is to select a fuse with an I^2t (ampere^2 second) rating below that of the device I^2t rating. The devices with the SC designations are General Electric triacs. Those with the C designations are General Electric SCRs.

CAPACITOR-DISCHARGE IGNITION SYSTEM

The capacitor-discharge system in Fig. 5-20 commands interest for at least three reasons. First, the operational mode is classical; that is, it is

TABLE 5-3. Fuse Selection Guide
for Phase-Controlled Lamp Circuits

Device	I^2t (amp² sec)
SC35/36	7.5
SC40/41	21.0
SC45/46	53.5
SC50/51	83.5
SC60/61	565.0
SC141	83.5
SC146	83.5
C106	0.5
C45/46	2000.0
C50/52	4000.0

Courtesy General Electric Co.

not likely to evolve into something drastically different. Second, it allows retention of the ignition breaker points, standard on all cars before the advent of factory-installed electronic ignition systems. Third, it provides the driver the *option* of selecting either the conventional ignition system or this electronic ignition system. This is psychologically desirable, because most of us tend to feel more secure with such an emergency provision.

It should be pointed out that the often-heard argument that the breaker points must be eliminated to have a truly electronic ignition system may be valid from the viewpoint of the purist but is not of overwhelming importance in practice. This is because once the breaker points have been relieved of the necessity to carry high current, their life span is appreciably increased. Moreover, the average motorist wants the basic benefits of electronic ignition in an existing car without the need for major surgery, such as would be involved in converting the distributor to use a magnetic reluctor or optoelectronics. The system described here is inexpensive, easy to install, and has good performance characterisitics.

In Fig. 5-20 power transistors Q1 and Q2, in conjunction with saturable transformer T1 and associated components, comprise an audio-frequency, square-wave inverter. The secondary voltage from this inverter is rectified by a bridge rectifier and filtered by capacitor C1. Approximately 175 volts dc is made available for operation of the energy charging circuit, the main components of which are capacitor C2, diodes CR1 and CR3, and inductor L1. The basic idea is to accumulate a charge in capacitor C2 quickly and, by means of SCR1, to "dump" this charge into the primary winding of the ignition coil. The timing of these events

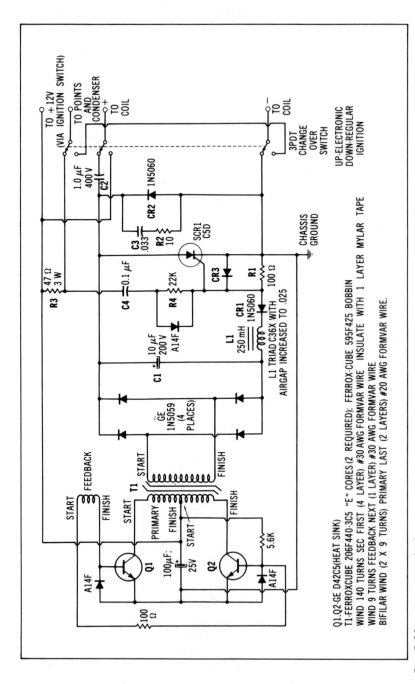

Fig. 5-20. Capacitor-discharge ignition system.

and the distribution to appropriate spark plugs occur in the same manner as in the conventional ignition system. The advantage of the capacitor-discharge ignition system over the older conventional (Kettering) type system is that a really respectable amount of electrical energy is available to fire worn, poorly adjusted, or fouled spark plugs. Of equal importance the spark-plug firing voltage not only holds up at high engine speeds with the capacitive-discharge ignition system, it is not readily depleted by low battery voltage when starting a cold engine.

This circuit is similar to most other capacitor-discharge ignition systems in that an SCR is triggered to transfer stored charge from a capacitor to the ignition coil. However, when we investigate the manner in which capacitor C2 is charged, a unique mode of operation becomes apparent. The significant deviations in this circuit are that only 175 volts are developed by the dc-to-dc converter capacitor C1, and the inclusion of inductor L1 and diode CR1. Initial appraisal might erroneously identify this as a low-energy system, because 175 volts and a 1-μF storage capacitor apparently do not compare favorably with other circuits. (This becomes particularly glaring when one contemplates that electrostatic energy storage in a capacitor is proportional to $V^2 \times C$.)

Actually the accumulation of stored energy is reinforced by so-called dc-resonance charging. This is a technique borrowed from radar modulators in which the voltage impressed across a storage capacitor is *double* the dc voltage of the power supply. When this system is in operation, a series-resonant circuit involving capacitor C2 and inductor L1 tend to produce shock-excited oscillations. Because of "stand-off" diode CR1, only the first quarter of oscillation can be sustained. However, this suffices to boost the 175 volts across capacitor C2 by another 175 volts, making the total potential across the terminals of C2 350 volts. Because of the way diode CR1 is polarized, the 350-volt capacitor charge is "trapped" until transferred to the ignition coil when SCR1 is triggered. Triggering of SCR1 is brought about in the following manner. When the ignition breaker points are closed, no path is available from the positive battery terminal to produce the required gate current for triggering. When the breaker points open, the path via resistor R3 is immediately available to charge capacitor C4 and thereby develop a pulse of gate current for the SCR. Subsequent closure of the breaker points then dissipates the charge in C4. The salient feature of this triggering technique is that the ignition breaker points, condenser, and ignition coil can be left as they were in the original system.

When the ignition breaker points open, the SCR triggers and remains conductive long enough to fire a spark plug. When turned on, the SCR is part of a resonant circuit formed by the primary inductance of the ignition coil and capacitor C2. At the outset of the first reverse-voltage alternation of the shocked oscillation, the SCR is turned off. Thus, the circuit is self-commutating. Both commutation and the described resonant charging for this capacitive-discharge ignition system work reliably over the entire range of engine speeds in 4-, 6-, and 8-cylinder automobiles.

PULSER FOR SOLID-STATE INJECTION LASERS

Fig. 5-21 shows a system for pulsing the solid-state injection laser with high currents at a low duty cycle. Not only is the circuit strongly reminiscent of gas-laser technology, but it will be seen that high voltage, rather than the several volts ordinarily involved with diodes, is used.

The two series-connected 40852 power transistors provide the path for "quick-charging" capacitor C *between* laser pulses. Although it might initially appear otherwise, the laser diode derives its pulsed energy from the charge stored in capacitor C, *not* directly from the power transistors. Therefore, the capacity of capacitor C determines the *duration* of the current pulse that activates the laser diode.

The 1N914 diodes are connected in the circuit in such a way that the lower 40852 transistor is deprived of forward bias while current is being delivered to the laser diode. But after depletion of the stored charge in capacitor C, the base of this transistor recovers its forward bias, and the charging process commences again. Inasmuch as the SCR in series with the laser diode may be pulsed at an audio-frequency rate, capacitor C must be capable of being charged rapidly to provide energy for the laser diode each time that the SCR is turned on.

Because the energy available for the laser diode is proportional to the square of the voltage stored in capacitor C, considerable control of the power residing in the laser beam is obtained from a relatively small variation of the voltage *allowed* to develop across the terminals of capacitor C. This adjustment is provided by the 250K potentiometer.

The capacitor voltage can only attain approximately the voltage at the wiper arm of the potentiometer because the conduction of the 1N3563 diode reduces the forward bias at the base of the output transistor. The charging rate of the capacitor is limited by this action so that

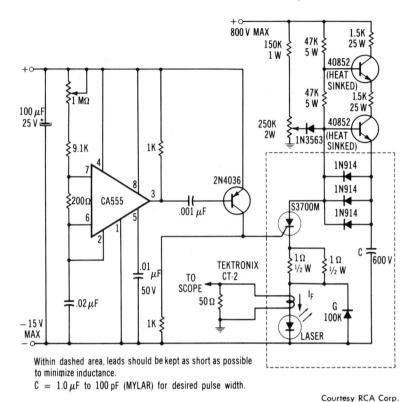

Within dashed area, leads should be kept as short as possible to minimize inductance.

C = 1.0 μF to 100 pF (MYLAR) for desired pulse width.

Courtesy RCA Corp.

Fig. 5-21. Pulser for injection laser devices.

the capacitor does charge to a voltage beyond that at the wiper arm of the potentiometer. When the wiper arm of the potentiometer is at its ground end, the capacitor voltage is substantially zero. At maximum adjustment of the potentiometer, the capacitor charges to 500 volts.

The SCR trigger pulses originate in the CA555 IC timer, which is connected to operate as an astable oscillator. The repetition rate of the trigger pulses is adjustable by the 1-megohm potentiometer. The output from this square-wave generator is made into narrow, positive-going, unidirectional pulses in the unbiased amplifier stage employing the 2N4036 transistor. Such pulses are more suitable for triggering the SCR than is the original signal.

Typical operating conditions of a family of RCA GaAs injection lasers are given in Table 5-4. The threshold current, I_{th}, is the current at which the device begins *lasing*. At lower currents the behavior is

TABLE 5-4. Typical Operating Conditions for Some RCA Injection Lasers

Characteristics at T_C = 27° C and at the Specified Operating Conditions							
Total Peak Radiant Flux (Φ_M) (Power Output) Minimum—W	Peak Forward Current (i_F) A	Pulse Duration (t_w) μs	Pulse Repetition Rate (prr) kHz	Typical Threshold Current (i_{th}) A	Typical Peak Forward Drop at i_F (v_F) V	Typical Rise Time ns	Type
1	10	0.2	1	4	4.5	<1	SG2001
2	10	0.2	1	3.5	4.5	<1	SG2002
3	25	0.2	1	6	6.5	<1	SG2003
5	25	0.2	1	6	6.5	<1	SG2004
5	20	0.2	1	6	6	<1	SG2005
7	40	0.2	1	11	8	<1	SG2006
10	40	0.2	1	11	8	<1	SG2007
12	75	0.2	1	25	10	<1	SG2009
15	75	0.2	1	25	10	<1	SG2010
20	100	0.2	1	36	12	<1	SG2012

essentially that of an infrared-emitting diode. Lasing action is characterized by certain desirable optical phenomena, coherent electromagnetic radiation at an extremely narrow bandwidth, and high power density in the beam.

Other solid-state optical devices can be operated in this circuit, including infrared-emitting diodes and LEDs. (A reduction of the high voltage would generally be necessary.) Also there are stacked-diode arrangements and diode arrays that can be used with this circuit. Such structures produce more beam power than it is practically feasible to obtain from a single laser diode.

WARNING

Laser radiation   laser devices in operation produce invisible electromagnetic radiation, which may be harmful to the human eye.

INDUCTION RANGE

Two modes of electrical cooking have enjoyed considerable success in both homes and restaurants. The oldest of these methods is the simple electric range. With this method of cooking, the rather straightforward technique of dissipative heat production in a resistive element is employed. A newer and more sophisticated method is found in the "radar" or microwave range. Here the heat needed for the cooking process is generated within the food being processed. The temperature rise in the affected food material is the result of dielectric losses and generally requires a minimal moisture content.

With the advent of suitable solid-state power devices, a *third* method of cooking with electricity has become economically and technically feasible. This is *induction* cooking and involves the production of eddy currents in a metal pot or cooking vessel. The cooking vessel functions essentially as the shorted secondary of a high-frequency transformer. The frequency has to be high relative to the 60-Hz power line frequency in order to produce sufficient heating effect, because the equivalent transformer has an "air core." If the vessel is made of iron or steel, there is additional heating because of magnetic hysteresis. However, best heating stems from the large eddy-current loss in copper vessels.

The "burner" of such an inductive range is the inductive element of a series-resonant *tank circuit.* For the reasons just stated, and for various other practical reasons, the resonant frequency is usually in the vicinity of 25 kHz to 40 kHz. The technique is both electrically and thermally efficient. The "burner," of course, remains relatively cool.

While such a cooking method seemingly has much merit, it has not yet become a commercial success. This may be due to the fact that inductive cooking has appeared on the scene just when the long-heralded impact of microwave cooking was finally materializing. In any event the circuit of the experimental induction range shown in Fig. 5-22 is bound to stimulate interest.

The heart of the circuit is the SCR series inverter consisting of inductor L1 (the "burner"), resonating capacitor C4, SCR2, diode RD2, and snubbing network R4 and C5. Inductor L1 is spirally wound with several turns of copper tubing to facilitate coupling to the cooking pot. A practical form of this inductor is readily made to have an inductance of about 6 μH. In conjunction with resonating capacitor C4, energy oscillates in these reactive elements and is inductively transferred to the cooking vessel through the physical proximity between the inductor and

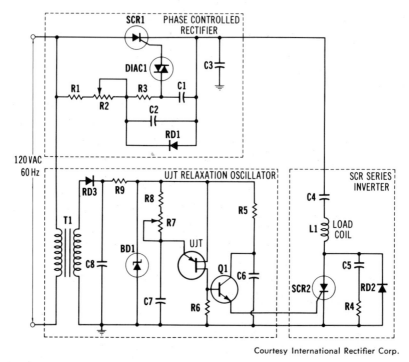

Courtesy International Rectifier Corp.

Fig. 5-22. Experimental circuit for an induction range.

the vessel. As might be suspected, SCR2 functions as a switch to produce shock excitation of the LC circuit. The oscillatory waveshape is *not* continuous, nor can it be identified as a *damped* waveform. This statement may appear to be contradictory, for most high-frequency electronic apparatus produce *either* continuous or damped oscillations. The operating mode of the inductive range is representative of a third way of "packaging" electrical energy, which is by bursts of integral cycles. Mathematically it can be shown that such waves are not damped. This is easily confirmed from a practical standpoint, because such waves do not produce harmonics, which are characteristic of all damped wave phenomena. The induction range produces a wave package consisting of *single* complete sine waves. This unusual waveform is shown in Fig. 5-23.

Assume that SCR2 has just been triggered to its on-state. The positive excursion of sinusoidal current then takes place in the series-resonant LC circuit. Ordinarily such a shock-excited LC circuit tends to "ring," that is, generate a decaying wavetrain of oscillatory waves. When the current attempts to reverse in this circuit, the SCR commutates or turns off. However, the reverse current finds a path through diode RD2. Further oscillation is prevented by the off status of the SCR. Thus, only *one* complete sine-wave cycle has been allowed to circulate in the series tank circuit. (To pacify the "purist," it must be conceded that a number of nonideal circuit situations exist and the integrity of the single wave falls short of perfection. For most practical purposes, however, it is satisfactory to identify it as a single sinusoidal wave of current.)

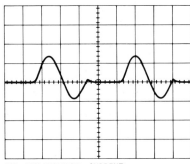

SCR2 AND DIODE RD2 CURRENT
50 AMPS/DIVISION
10μs/DIVISION

Fig. 5-23. Waveforms produced by the induction range.

Courtesy International Rectifier Corp.

The gate of SCR2 is repetitively triggered at approximately 22 kHz, or at a frequency about 70 percent of the resonant frequency of the series LC circuit. That this is so can be readily deduced by comparing the dead time in Fig. 5-23 with the time required for a complete cycle. This mode of operation is suitable for the SCR, for it then has more than sufficient time to turn off between cycles.

The triggering circuit is a conventional unijunction transistor relaxation oscillator with an emitter follower as a buffered output. This circuit is powered by a simple half-wave power supply working from a 25-volt isolation transformer.

The dc power supply for the induction range is designed around SCR1, which performs essentially as a controlled output half-wave rectifier. Except for the filter capacitor connected at its output, this circuit is similar to that used in lamp dimmers and in motor speed controls in electric tools. The extra components associated with the gate circuit are intended to reduce the hysteresis that can often be a nuisance with "economy" light dimmers. Such a gate circuit is known as a *double-time-constant type*. It enables continuous control settings from near zero to maximum output. The rather primitively filtered dc is entirely suitable for shock excitation of the LC resonant circuit. (Of course, the experimenter might find it more convenient to use a large variac and a bridge rectifier for the power supply.)

OPTICALLY TRIGGERED HIGH-VOLTAGE SWITCH

Performing switching operations at high voltages is not always the trivial task that we tend to associate with such a "simple" function. Often the requirements are speedy actuation, electrical isolation between the high-voltage circuit and the control circuitry, immunity to false actuation from electrical interference, and "bounceless" closure of the high-voltage circuit. Inasmuch as the switching action is often intended to initiate a precise pulse of energy, electromechanical control techniques are inherently fraught with difficulties. The enthusiastic motivation to do things electronically is often modified by techniques that *combine* electronics with optical technology. The high-voltage switch shown in Fig. 5-24 is a unique example of this evolving design approach.

This high-voltage switch has a myriad of applications such as in lasers, masers, electron microscopes, X-ray equipment, stroboscopy,

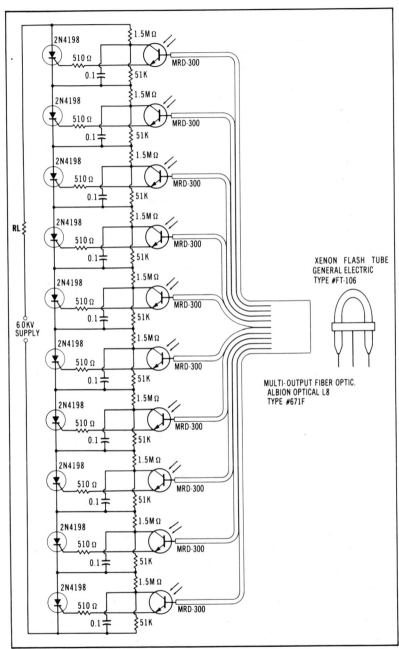

Fig. 5-24.   Optically triggered high-voltage switch.

particle accelerators, copying machines, and power supplies. It is a "natural" as a crowbar circuit for protective shorting of a high-voltage source. Although the SCRs tend to latch in their on state, this does not happen in crowbar applications where the voltage of the shorted source is reduced below that needed to supply holding current to the SCRs. Thus, this crowbar action can be self-clearing if the optical triggering signal is momentary, rather than sustained.

The MRD-300 phototransistors and the fiber-optical system constitute the heart of this switching scheme. For all practical matters it can be stated that these transistors simultaneously receive their optical turn-on signals regardless of the length of the individual light fibers. When these transistors are turned on, they transfer the stored charges from the 0.1-μF capacitors into the gate circuits of the respective SCRs.

The turn-on time of this high-voltage switch is on the order of 300 nanoseconds at a 1-ampere load current. In pulse applications it can supply 100 amperes to the load for a duration of 4 milliseconds. For best results the SCRs should be selected to have closely matched rise times.

It is anticipated that this method of triggering series-connected thyristors will prove superior to the older technique of providing each thyristor with its individual pulse transformer.

SIMPLE HALF-WAVE UNIVERSAL
MOTOR CONTROL CIRCUIT

The single SCR phase control circuit shown in Fig. 5-25 is both simple and economical. Nonetheless such a primitive arrangement will serve many purposes. The neon bulb is one with capability for handling high current pulses.

When capacitor C attains sufficient voltage to ionize the neon bulb, its charge is abruptly depleted as a trigger pulse is delivered to the gate of the SCR. This happens only on the positive excursion of the ac line voltage. During the negative half-cycle the residual voltage in the capacitor is further reduced, and both the neon bulb and the SCR gate circuit appear as open circuits. The speed with which the capacitor voltage can build up to the value needed to fire the neon bulb depends on the setting of potentiometer R2. Therefore, R2 controls the timing or phase of the trigger pulse relative to the power-line voltage wave.

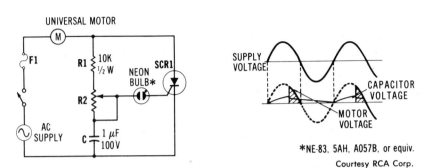

Fig. 5-25. Simple SCR universal motor control circuit.

This is tantamount to saying that R2 provides adjustment of the conduction time of the SCR and therefore of the average current allowed to pass through the universal motor. In this way the speed of the motor is manually controllable.

This simple circuit does not feature wide-range speed control. The SCR conduction angle may be varied from only about 30 to 150 degrees. Even if full 180-degree conduction of the SCR could be attained, the motor speed would be limited to about 75 percent of its line-voltage speed because of the half-wave operation. Table 5-5 lists the circuit components in Fig. 5-25 for various operating conditions.

TABLE 5-5.   Components for Circuit Shown in Fig. 5-25

AC Supply	AC Current	F_1	CR_1	R_2	SCR_1
120 V	1 A	3 AG, 1.5 A, Quick Act	D1201B	100 K, ½ W	RCA-2N3528
120 V	3 A	3AB, 3 A	D1201B	100 K, ½ W	RCA-2N3228
120 V	7 A	3 AB, 7 A	D1201B	100 K, ½ W	RCA-2N3669
240 V	1 A	3 AG, 1.5 A, Quick Act	D1201D	150 K, ½ W	RCA-2N3529
240 V	3 A	3AB, 3 A	D1201D	150 K, ½ W	RCA-2N3525
240 V	7 A	3 AB, 7 A	D1201D	150 K, ½ W	RCA-2N3670

Courtesy RCA Corp.

HALF-WAVE CONTROLLER WITH SPEED REGULATION

The SCR circuit shown in Fig. 5-26 is also intended for universal motors. Although rivaling the control scheme of Fig. 5-25 in simplicity, this

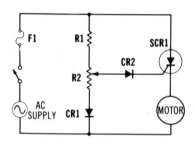 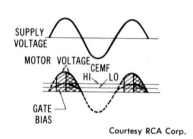

Courtesy RCA Corp.

Fig. 5-26. Speed-regulated controller for universal motors.

controller provides *speed regulation*. Inasmuch as there are no apparent feedback loops, this may appear surprising. Speed regulation is attained in the following way.

While the SCR is blocking, residual magnetism in the motor causes a counter emf to appear across the motor terminals. This counter emf is proportional to the rotational speed of the armature. In order for the SCR to turn on, the gate voltage at the wiper arm of potentiometer R2 must exceed the counter emf produced by the motor, the voltage drop across diode CR_2 and whatever gate-cathode voltage is needed for triggering. Because this is a half-wave circuit, the SCR can turn on only during the positive half-cycles.

Suppose that the motor is operating at some nominal speed and its mechanical load is increased. Under such conditions it is the nature of the universal motor to slow down. However, in this circuit a reduction in the speed of the motor results in decreased counter emf. This in turn enables the SCR to fire earlier in the positive half-cycle of the power-line voltage wave. As a result the average motor current increases, thereby causing the motor to increase its speed. The reverse sequence of events applies if the load on the motor is relaxed and the motor attempts to increase its speed. The overall result is that good speed regulation is achieved. The speed regulation is operative not only for variations in shaft loading but for other disturbances such as erratic brush behavior, varying bearing friction, and temperature effects.

In a circuit of this type, where no RC phase-shifting network is used, the phase delay of the trigger pulse is limited to 90 degrees. In actual applications the motor cannot be reliably controlled at slow speeds and light loads. It can be appreciated that there is no counter emf at zero speed. This causes the SCR to fire early in the cycle, thereby accelerating the motor to a speed beyond that corresponding to the setting of R2. The motor must then give up its excess speed, via windage and friction. When it has slowed considerably, the SCR is able to fire again, only to impart another such impulse to the motor. When R2 is set for somewhat higher speed *and* if the motor is a reasonably loaded, such erratic operation, known as *skip-cycling,* cannot occur. This is because speed-ups and slow-downs then occur within the "capture range" of the feedback control and the speed is essentially stabilized.

Table 5-6 lists the circuit components used in Fig. 5-26 for various operating conditions.

TABLE 5-6.   Components for Circuit Shown in Fig. 5-26

AC Supply	AC Current	F_1
120 V	1 A	3 AG, 1.5 A, Quick Act
120 V	3 A	3 AB, 3 A
120 V	7 A	3 AB, 7 A
240 V	1 A	3 AG, 1.5 A, Quick Act
240 V	3 A	3 AB, 3 A
240 V	7 A	3 AB, 7 A

CR_1, CR_2	R_1	R_2	SCR_1
D1201B	5.6 K, 2 W	1 K, 2 W	RCA-2N3528
D1201B	5.6 K, 2 W	1 K, 2 W	RCA-2N3228
D1201B	2.7 K, 4 W	500, 2 W	RCA-2N3669
D1201D	10 K, 5 W	1 K, 2 W	RCA-2N3529
D1201D	10 K, 5 W	1 K, 2 W	RCA-2N3525
D1201D	5.6 K, 7.5 W	500, 2 W	RCA-2N3670

Courtesy RCA Corp.

MOTOR START SWITCH FOR $^1/_2$-HP CAPACITOR-START INDUCTION MOTOR

The compelling feature of the induction motor is that its construction is basically simple and it *runs* without the need for brushes or commutators. Unfortunately the much-used single-phase induction motor is not self-starting. Several starting techniques are employed. One of them connects a capacitor in series with a *starting winding* until the motor attains about 75 percent of full-load speed, at which time the starting circuit is interrupted. This is brought about automatically by means of a shaft-mounted centrifugal switch. Obviously such an electromechanical device increases the maintenance requirements. The centrifugal switch actually reduces the extremely good long-time reliability that otherwise would be expected from this motor.

Fig. 5-27 shows a scheme for dispensing with the centrifugal switch in induction motors. A $\frac{1}{2}$-horsepower induction motor is automatically taken through its start and run operating modes via a triac. The gate of the triac is connected to the sensing resistance, R1. In this application, resistance R1 is selected to provide gate-triggering voltage for motor currents exceeding 12 amperes rms. For this motor the *peak* starting current is in the vicinity of 40 amperes and the *peak* running current is about 8 amperes. (We are concerned here with *peak,* rather than rms, values because the gate of the triac responds to the peak value of the voltage drop sensed across R1.) It was also determined that the maximum starting current occurred about 12 cycles after application of the 60-Hz voltage from the power line. Such start-up characteristics for the motor is with the motor carrying its rated mechanical load.

When the ac line is connected to the motor, the large inrush of current develops a sufficient voltage drop across R1 to trigger the triac into its fully on state of conduction. Under this condition, starting capacitor C2, the starting winding, and the power line are connected in series. Because of the action of capacitor C2, the currents in the starting and running windings are displaced nearly 90 degrees in phase from one another. The motor then behaves as a *two-phase* machine and accordingly develops starting torque. (A simple single-phase field simultaneously urges the stationary rotor to turn in both directions and therefore produces no starting torque.) As the motor gathers speed, the line

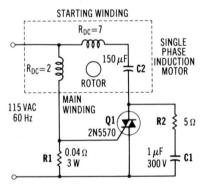

Courtesy Motorola Semiconductor Products, Inc.

Fig. 5-27. Motor start switch for $\frac{1}{2}$-HP capacitor-start induction motor.

current diminishes until the triac no longer receives adequate trigger voltage to keep it turned on. As mentioned, the starting circuit is opened at about 75 percent of full-load running speed and the motor thereafter carries its load as a true single-phase machine.

Because of the inductive nature of the starting winding, the snubber resistor than is ordinarily employed with noninductive loads.

FULL-WAVE SPEED-CONTROL CIRCUIT FOR ELECTRIC DRILLS

The motor control circuit of Fig. 5-28 not only provides manual speed control but incorporates feedback that increases available torque when the load becomes greater. In an electric drill the motor naturally slows when encountering increased load in an effort to develop higher torque via greater line current. The demand for more current is met in this control circuit by automatically increasing the voltage applied to the motor.

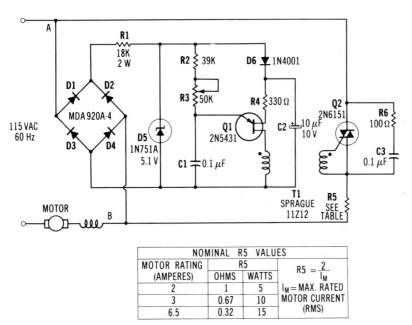

NOMINAL R5 VALUES			
MOTOR RATING (AMPERES)	R5		$R5 = \dfrac{2}{I_M}$
	OHMS	WATTS	
2	1	5	I_M = MAX. RATED
3	0.67	10	MOTOR CURRENT
6.5	0.32	15	(RMS)

Courtesy Motorola Semiconductor Products, Inc.

Fig. 5-28. Full-wave speed-control circuit for electric drills.

Inasmuch as separate access to the armature and field windings is not required, the application is simplified. In many instances it will be feasible to package the circuit within the tool or appliance itself. The circuit is similar to that of the basic unijunction trigger circuit shown in Fig. 3-22. The differences involve the addition of circuit components R5, D6, and C2. The objective of the additional components is to make the basic circuit operate so that the conduction angle of the triac is advanced as the motor demands more current. Speed regulation is *not* accomplished in this circuit because the sensed quantity, current, is not proportional to speed (as would be counter emf). However, there may be *some* boost in motor speed by virtue of the increased torque produced.

The bridge rectifier consisting of diodes D1, D2, D3, and D4 applies full-wave rectified voltage to the phase-control circuit. Phase control of the triac results from the charging of capacitor C1 through resistors R2 and R3 from the voltage level established by zener diode D5. When capacitor C1 develops sufficient voltage to fire unijunction transistor Q1, the triac receives a trigger pulse through transformer T1. Capacitor C1 quickly discharges through the emitter circuit of Q1.

During the time that the triac is *conducting,* the voltage between circuit points A and B is reduced to the extent that the voltage developed across capacitor C1 is not sufficient to fire Q1. The actual voltage *applied* to capacitor C1 then depends on the motor current because of the sensed voltage drop across sampling resistor R5. Specifically the greater the motor current, the greater will be the sensed voltage between points A and B and the more of a "head start" capacitor C1 will have when it commences its charging cycle to fire the unijunction transistor. In other words the triggering time of the triac will be advanced more as the motor encounters greater load and demands higher current. Resistor R5 is chosen so that the voltage applied to capacitor C1 during triac conduction cannot be great enough to break down zener diode D5.

The main difference from a conventional unijunction triggering circuit, the addition of resistor R5, has been accounted for. It is appropriate to explain the addition of diode D6 and capacitor C2. As pointed out, the amount of the preliminary charge imparted to capacitor C1 depends on the motor current. The net result is that the triggering of the triac is more and more advanced as the motor current becomes greater from heavier mechanical loading. This being so, it can readily be appreciated that Q1 must *not* retrigger while the triac is conducting. However, without diode D6 and capacitor C2, there would be a tendency for such malperformance. The reason for this is that the interbase voltage of the unijunction transistor would then be quite low, making it vulnerable to firing from low emitter voltage. In order to prevent this situation, the interbase voltage is maintained at a high level during triac conduction with the assistance of diode D6 and capacitor C2.

Successful operation has been demonstrated with 2- and 3-ampere motors used in $1/4$-inch drills. Under these conditions, motor speed has been satisfactorily controlled down to one-third, or less, of maximum speed, and the torque enhancement feature has proven its worth. Because of the varying characteristics of different motors, some experimentation may be in order to optimize the value of resistor R5. If this resistance is too high, a surging mode of operation will occur; if it is too low, full advantage will not be obtained from the torque boost.

REVERSING MOTOR-DRIVE CIRCUIT FOR
SERVOS AND OTHER APPLICATIONS

A circuit that enables the control of motor speed and direction rotation by means of a dc voltage obviously has many potential uses. Such a circuit is shown in Fig. 5-29. Its operation is based on the unique characteristics of the diac, triac, bridge rectifier, and a reversible series motor. Although these elements are often encountered in various other circuits, their utilization in this scheme is quite novel.

Consider first the relevant operating features of series motors in which the same current passes through the armature and field windings. This is why the series motor is capable of operation from an ac source; that is, the *relative* torque between armature and field is always exerted in the same direction despite polarity changes. By the same token the series motor can be reversed by changing the connections of either (but not both) the field winding or armature winding. It is only natural then that a so-called reversible series motor is one with independent access to the armature and field leads.

Now that we have a motor that lends itself readily to both speed control and reversal, it would be desirable to put it through its paces without resorting to mechanical switches. This is where the bridge rectifier circuit enters the picture. When the current source (the triac) provides one polarity, the current path through the bridge is through the pair of diodes appropriately polarized to be forward biased. Of course, current of opposite polarity completes its path through the other

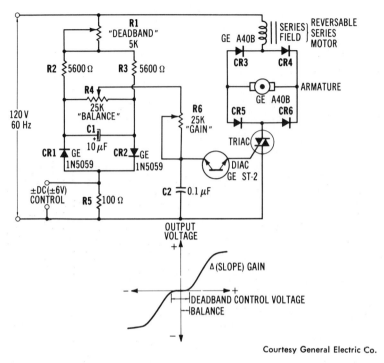

Fig. 5-29.   Reversing motor-drive circuit for servo systems and other applications.

pair of diodes. Specifically, positive current out of the triac goes through diode CR6, through the motor armature, through diode CR3, through the field winding, and back to the power line. Negative current from the triac goes through diode CR5, through the armature, through diode CR4, through the field winding, and back to the power line. Switching from one polarity to the other is the same as *interchanging* the armature connections of the motor.

From what has been said, it should be evident that the triac must *not* provide the usual full-wave current desirable in many control techniques. In this scheme it is already obvious that the triac must be able to provide *either* positive or negative current pulses but *not both*. If "normal" triac operation were used, the motor would be urged to turn one direction during the first half of the ac cycle and the opposite direction during the subsequent half-cycle. Over a full ac cycle the average torque would be zero, and much heat and noise would be forthcoming, but the motor shaft would not turn. Inasmuch as the triac can be triggered by gate signals of either polarity, it might appear difficult to make it deliver unidirectional current pulses at will. This problem, however, is neatly solved by the diac and its associated circuitry.

The diac is, of course, bilateral in that it will respond to either a positive or negative voltage of sufficient magnitude. If the diac is connected in a phase-shift circuit with a provision for mixing a dc voltage with the ac, the bilateral action of the diac circuit will be altered. Specifically the diac can be made to respond to *either,* but *not* both, the positive or negative half-cycle of the ac wave. This is what has been done in the diac circuit of Fig. 5-29. As a result of the dc control signal, capacitor C2 starts its charge cycles from a definite dc level rather than from near zero as in most triac control circuits. If, for example, capacitor C2 is initially several volts positive because of the dc control signal, the diac will be enabled to fire on positive half-cycles *only.* The reverse situation prevails with a negative dc control signal. Diodes CR1 and CR2 are "steering diodes" that provide the needed paths for the dc control signals but isolate the ac from the dc control terminal. The overall effect of this circuitry is that the triac is operated in the *half-wave,* rather than the more common full-wave, mode. (The control range and maximum speed limitations generally associated wih half-wave motor control systems, therefore, prevail with this scheme.)

Variable resistance R1 provides adjustment of the ac voltage available to the phase-control network. By reduction of this voltage a deadband is imposed in the control characteristic. This is necessary to ensure that full-wave operation of the triac will never occur. On the other hand the *smaller* the deadband, the more suitable the control is for servo operation. Balance control R4 adjusts the symmetry of the control characteristic with respect to a zero control signal. It compensates for whatever deviations from balanced control characteristics may accumulate from tolerances in the diac, triac, and motor. Variable resistance R6 is the counterpart of the phase adjustment in ordinary phase-control systems. In this circuit its effect is suggestive of a "gain" control inasmuch as the basic speed-control function is under command of the dc control signal. All three controls interact somewhat, and careful adjustment is initially required to establish desired operating conditions.

STATIC SWITCHES

The static switch is closely related to the solid-state relay but tends to be a much simpler circuit configuration. This is because the static switch usually does not have the additional components found in solid-state relays. Thus, the static switch generally will not incorporate opto-isolators, zero-crossing circuitry, output rectifiers for dc loads, or various logic provisions. However, static switches with considerable sophistication are coming into prominence, and it is no longer entirely clear whether *solid-state relay* or *static switch* is the proper nomenclature in some instances. The static switch is sometimes referred to as a *contactor*.

The best way to gain insight into the nature of solid-state switches is to investigate some typical applications. The simple triac static-switch circuit in Fig. 5-30 is widely used. Its main function is to turn on and off a high-current load by means of a small switch with a relatively low current rating.

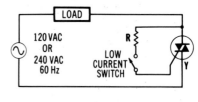

	120 VAC, 60 Hz	240 VAC 60 Hz
R	1K, ½ W	2K ½ W
Y	RCA 40429	RCA 40430

Courtesy RCA Corp.

Fig. 5-30. A simple triac static switch.

The circuit shown in Fig. 5-31 is the SCR equivalent of the triac static-switch in Fig. 5-30. Diode D2 provides triggering current for the gate of SCR Q1, and diode D1 serves the same function for the gate of SCR Q2 during the alternate half-cycle. Resistor R1 limits the gate current to both SCRs.

Three other simple static-switch circuits are shown in Fig. 5-32. The scheme depicted in Fig. 5-32A is similar to the triac circuit already considered in Fig. 5-30, except that a reed relay is used to switch the gate current. This minor expedient provides a worthwhile feature—the

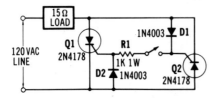

Fig. 5-31. Static-switch circuit using two SCRs.

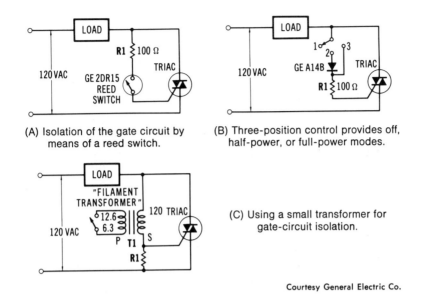

(A) Isolation of the gate circuit by means of a reed switch.

(B) Three-position control provides off, half-power, or full-power modes.

(C) Using a small transformer for gate-circuit isolation.

Courtesy General Electric Co.

Fig. 5-32. Additional triac static-switch circuits.

electrical isolation of the gate circuit. Such isolation is desirable in terms of safety and control flexibility. Otherwise, grounding conflicts can produce various malfunctions and failure modes. The three-position control circuit in Fig. 5-32B is interesting in that the diode in position 2 causes the triac to deliver only half-wave pulses to the load. The circuit action is the same as if one of the SCRs in Fig. 5-31 were removed. Thus, the triac is caused to operate as an SCR, and the load power in position 2 is one-half of that obtained in position 3.

The circuit shown in Fig. 5-32C is an unusual method for producing on-off load control with gate-circuit isolation. When the switch is open, the secondary winding of the transformer (in this case the 120-volt winding) has a high impedance and therefore does not allow sufficient gate current to fire the triac. However, when the switch is closed, the secondary winding "sees" the reflected short circuit of the primary winding. The impedance of the secondary winding is therefore low, and there is sufficient gate current to turn on the triac.

For all of the simple static switches described here, some experimentation may be in order to determine the optimum value of the gate-current limiting resistance. This resistance will be influenced by the characteristics of the particular thyristor used, by temperature, by line-voltage variation, and by load impedance. For the most part, these simple circuits operate best with essentially resistive loads.

A more sophisticated static-switching system is shown in Fig. 5-33. This circuit permits application and interruption of three-phase power to an induction motor or to appropriately connected and rated lamps or heater elements. This arrangement features single logic-signal control of the three-phase currents and good isolation between the logic control circuitry and the ac power line. The isolation is provided by the photo-coupled isolators. The three CA3059 ICs are employed here as differential op amps and deliver sustained dc turn-on signals to the logic triacs when the dc input logic is high. The logic triacs function essentially as driver stages for the power triacs, which actually perform the load-current switching functions.

The ICs have internal dc power supplies, but external filtering is required. This is provided by the three 100-μF capacitors connected between terminals 2 and 8 of the ICs. Although these ICs can provide zero-voltage switching for rfi reduction, they are not connected here to take advantage of this feature. When an inductive load such as a motor is switched, the inductance slows down the rate of load current rise, thereby greatly reducing the chief source of rfi in thyristors. On the other hand the switching of resistive loads, such as lamps or heaters, *will* be accompanied by bursts of rfi. Inasmuch as the switching will generally be performed at widely spaced intervals, it is reasonable to assume that rfi will not be a problem in many applications.

The terminal labeled *ref* is an artificial "neutral" established by the 8K resistors for the three-phase power line. This provides proper sequencing of the ac operating power for the three ICs. A 220-volt power line is not required because these ICs are specified for essentially constant characteristics over an ac operating voltage range of 120 to 277 volts.

LIMITED-RANGE THREE-PHASE
CONTROL OF LOAD POWER

At first the circuit in Fig. 5-34 appears to be a scheme for amplitude control of the triac gate in a way somewhat analogous to the method that one might use to control transistor load current by varying the base current. It is true that variable resistance R2 controls the current in the gate circuit of the triac. However, because the triac is either on or off,

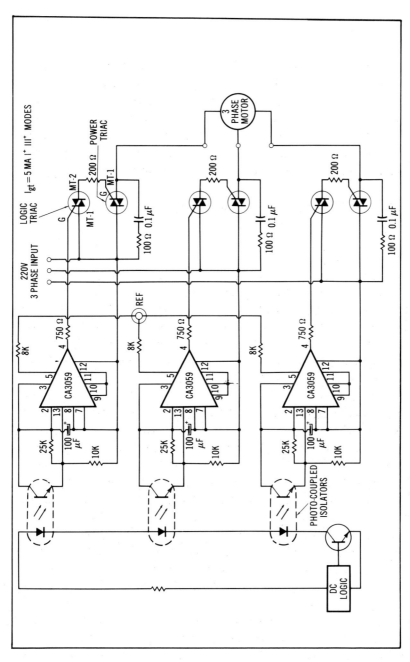

Fig. 5-33. Triac contactor circuit for three-phase loads.

Fig. 5-34. Simple control circuit for applications where limited control range is allowable.

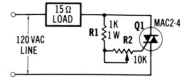

Courtesy Motorola Semiconductor Products, Inc.

the load current is not varied by adjustment of the gate current. What actually happens is that resistance R2 varies the time required for the applied gate voltage to attain triggering level. Thus, the circuit in Fig. 5-34 is a primitive phase-control technique. However, unlike the more common triggering arrangements utilizing RC circuits and triggering diodes, this scheme can provide load current control only between 90 and 180 degrees of each line-voltage alternation. Resistance R1 limits the gate current, and its value should be experimentally determined for different triacs. This method of load control is commendable in terms of economy and simplicity for those applications where limited control range is not a disadvantage.

USE OF A TRIAC TO PREVENT
CONTACT ARCING IN A RELAY

Electromechanical relays have the desirable feature that the voltage drop across the closed contacts is a small fraction of that attainable with solid-state devices. However, arcing tends to shorten the life of the contact material, and objectionable rfi is often produced in the process. Also contact bounce sometimes causes additional problems, especially when there are logic circuits in the system. By combining a relay with a triac, the desirable features of *both* devices can be realized in many instances. The cost of the combination can actually be *less* than that for one device alone because the current rating of the relay contacts can be drastically reduced. Also a heat sink that might be required for a triac alone is not likely to be needed for the relay-triac combination.

The basic idea is to have the triac turn on *before* the relay contacts close and to remain on *after* the relay contacts open. While the relay contacts are closed, the triac is shorted out, thereby developing negligible voltage drop and power dissipation. A circuit arrangement for accomplishing this logic is shown in Fig. 5-35. The main current-carrying

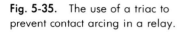
Fig. 5-35. The use of a triac to prevent contact arcing in a relay.

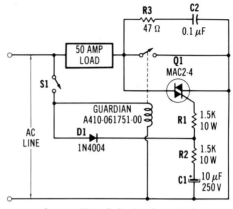

Courtesy Motorola Semiconductor Products, Inc.

terminals of the triac are connected directly across the relay contacts. When switch S1 is closed, the triac is immediately triggered through diode D1. The relay does not actuate immediately because of its solenoid inductance and its mechanical inertia. When the relay finally does respond, its contacts then take over the task of conducting the load current. At the same time capacitor C1 accumulates a charge. When switch S1 is opened, the inherent lags in the relay again prevent immediate response. When the relay contacts finally open, the triac *remains* turned on because of the charge stored in capacitor C1. When the charge depletes itself in the gate circuit, the triac finally turns off. In this sequence of events the triac turned on *before* the closure of the relay contacts and remained on *after* the opening of the relay contacts. Therefore, the maximum voltage developed across the relay-triac combination is on the order of 1.5 volts. Also the load is virtually spared the effects of contact bounce.

The closing and opening time lags for the relay specified in Fig. 5-35 are each about 15 milliseconds. With other relays, some experimentation may be necessary for the values of R2 or C1. The snubber network, R3 and C2, protects the triac from possible ac-line transients. Because there is no welding or pitting of the contacts, a relay can be selected with one-fifth to one-tenth the current capability that would otherwise be mandatory to provide a good safety margin for the effects of contact arcing.

By replacing switch S1 with a photo-isolator device, isolation from the line voltage is provided and remote actuation is made more feasible.

(Such a system would be useful for many electrical control applications but appears to have been largely overlooked by the manufacturers of solid-state relays. This has apparently been because the mixing of semi-conductor and electromechanical components violates the concept of a "pure" solid-state relay.) The opto-isolator circuitry of Figs. 5-3 and 5-33 should prove useful for this purpose.

6

From the Classic to the Avant-Garde: A Look at Newer Developments

The power devices and circuit implementations discussed in the previous chapters represent accepted design practice for a decade, and often longer. Indeed much of electronic control of power appears as classic in nature as one would dare hope in the dynamic technology of semiconductor devices. To be sure, improved devices are continually introduced on the market—a power transistor, for example, now displays less leakage, develops a higher current-gain, has greater electrical ruggedness and reliability, and is a more uniform product offered at a lower cost than its predecessor of a few years ago. Nonetheless great leaps forward do not generally revolutionize design approaches on an annual basis.

It is the objective of this chapter to exemplify those unique advances in devices and in circuitry that appear as harbingers of the state of the art for the immediate future. Most of these advanced devices are well out of the laboratory; others are just rising on the horizon. In any event a change that is more in the nature of natural evolution than abrupt mutation is taking place (such as experienced when semiconductors began impacting the domain of vacuum tubes). Former and present solid-state power techniques will not necessarily be rendered obsolete—at least, not quickly so. It is clear, however, that a wider range of choices and greatly extended flexibility will enable the designer to control power in interesting new ways; size, weight, cost, versatility, and efficiency will be more readily optimized than hitherto.

The newer devices and applications are discussed in potpourri sequence, it being assumed that the previous chapters have served to illustrate the general ideas underlying solid-state power electronics. With this editorial slant the avant-garde devices and circuitries make their debut in a timely manner. Some of the newer devices are characterized by sophistications not previously attainable on a practical basis; other devices qualify via their obviously extended performance over what we have been conditioned to expect in next year's model. Even as this is written, the impact of solid-state control is being dramatically extended in automotive, utility, rf and microwave, space-vehicle, and other fields. The performance of electric motors is increasingly being dictated more by combined logic and solid-state control than by the textbook-ascribed technical "personalities" of the motors themselves. All in all, solid-state control of electrical power is destined to become one of the dominant features of our industrial civilization.

THE SYNCHRONOUS RECTIFIER

Sometimes it is hard to say which is the cart and which is the horse when it comes to useful technologies; devices and circuits beget one another. A useful circuit is a natural challenge for developing advanced devices in order to boost the performance of the circuit. But it is likewise true that new circuits tend to form in response to the introduction of better devices. At this writing, the synchronous rectifying circuit is being re-introduced to the technical community because of anticipation that recent, and especially, forthcoming power MOSFETs can make optimal use of this not-so-new circuit. It will serve both historical interest and technical relevancy to uncover the recent past of this novel method of producing rectification.

First of all, the synchronous rectifier is an *active* circuit for converting alternating current to direct current—it uses active devices, such as transistors, rather than the traditional passive diode. Unilateral conduction (rectification) is achieved by applying an appropriately timed signal to the control element of the devices. Thus, the basic principle is simple and straightforward. A natural question, however, is, Why do it in this manner when the use of rectifying diodes is even simpler? The answer is that under certain circumstances it is possible to obtain greater rectifying efficiency than with rectifying diodes. Designers go great lengths to squeeze a few percentage points more overall efficiency from converters and regulated power supplies. The rewards in the marketplace can justify every percentage point of needless dissipation eliminated.

A number of conflicting factors have been associated with synchronous rectification. At one time Motorola advocated the use of this circuit technique because it marketed germanium power transistors that could be incorporated in a power supply and yield a voltage drop of only 0.3 volt when supplying 60 amperes of direct current to a load. (See Fig. 6-1.) This is much better than can be obtained with diffused silicon rectifying diodes and is competitive with the performance of Schottky diodes. Several things happened to dampen enthusiasm over the germanium transistor synchronous rectifier. Motorola and other large manufacturers of germanium power transistors abruptly terminated this activity because of the manifold advantages of silicon for *most* applications. Also switching power supplies moved from low-audio switching rates to the vicinity of 20 kHz. At this higher frequency the switching losses in germanium transistors would overwhelm the small loss resulting from the low voltage drop. Finally, the Schottky diode developed rapidly; it provided low voltage drop, too, and could efficiently handle much higher frequencies than 20 kHz. Because of other limitations of the Schottky rectifier—its low voltage rating and its high-leakage current—designers remained intrigued with the idea of synchronous rectification.

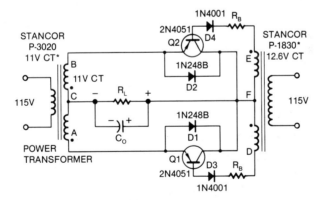

Fig. 6-1. Synchronous rectifier for high-current, low-voltage loads. The germanium power transistors develop low voltage drop. Rectifying efficiency can therefore exceed that attainable with conventional rectifying diodes.

The MOSFET Synchronous Rectifier

Before describing a MOSFET synchronous rectifier circuit, it is well to recall that one of the disadvantages long cited as inherent in the MOSFET device is its relatively high on resistance. However, technology has been advancing so that the resistance of large MOSFETs is already low enough to merit its consideration as a synchronous rectifier. Moreover, manufacturers are busy developing dedicated types that should have such low on resistances that their use in synchronous rectifiers will be compelling to circuit designers.

The MOSFET synchronous rectifier shown in Fig. 6-2 is, like the germanium transistor circuit, a full-wave configuration. Instead of utilizing a center-tapped transformer, the circuit (Fig. 6-2) makes use of the popular four-element bridge connection. Although the low parts-count catches the eye, this circuit is even simpler than a superficial inpection suggests. This stems from the fact that the diodes shown are internally part and parcel of the MOSFET structure. Indeed all power MOSFETs have these "parasitic" diodes. In many applications the diodes are of no consequence. Sometimes their presence cannot be ignored, but any adverse effects can be circumvented by appropriate circuit design. In still other applications the internal diode can be *advantageously* used, thereby eliminating the need for a discrete diode. Examples are often found in various motor control and regulator circuits, where a

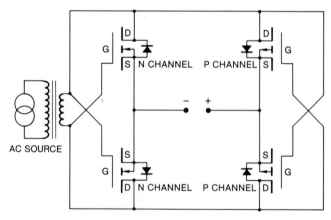

Courtesy Siliconix Inc.

Fig. 6-2. Synchronous bridge rectifier using power MOSFETs. The diodes are not discrete devices but rather intrinsic elements common to all MOSFETs.

diode is required for such functions as free-wheeling dynamic braking, or regeneration. As will be seen, it is necessary to discuss this diode with regard to its effect on the synchronous rectifier circuit. Interestingly at least one manufacturer often uses the MOSFET symbols depicted in Fig. 6-3.

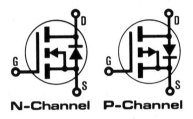

N-Channel P-Channel

Courtesy International Rectifier Corp.

Fig. 6-3. Symbols of international rectifier's HEXFET devices. The diodes are internal—intrinsic in the fabrication of all power MOSFET devices. In some circuits these diodes are of no consequence, in others their presence can adversely affect operation. In still other applications the internal diode is put to good use, thereby dispensing with a discrete diode.

In the ensuing description of this synchronous rectifier circuit, it will be seen that a MOSFET property that has no analogy in either tubes or bipolar transistors is exploited. Specifically the MOSFET will operate with either polarity on its "plate" or "collector," i.e., on its drain. If we liken an n-channel MOSFET to a tube, such a tube would have to be able to operate with negative voltage on its plate. Similarly an analogous npn bipolar transistor would have to operate with negative voltage on its collector. It is also true that the p-channel MOSFET, although usually operated with negative drain voltage, is capable of operation with positive drain voltage as well. This unique behavior makes the MOSFET a more versatile device than its "analogous" counterparts.

Despite the circuit simplicity of the MOSFET synchronous rectifier of Fig. 6-2, its operating mode is not immediately apparent. In the following explanation, keep in mind the three things previously alluded to:

- The germanium synchronous rectifying circuit of Fig. 6-1 *simulates* the rectifying action of diodes because the transistors are turned on and off at appropriate intervals of the ac cycle. This basic idea pertains to the MOSFET circuit also.

- MOSFET power transistors have internal drain-source diodes as depicted in Fig. 6-3.
- MOSFET power transistors can conduct with *either* polarity of voltage applied to the drain.

Consider now the upper-left MOSFET of Fig. 6-2. Assume that the drain-source voltage is negative—*opposite* to what it would be in "conventional" circuits, such as amplifiers. The gate will simultaneously be positive, thereby making this transistor conductive. One might argue at this point that the transistor will be shorted out by its internal drain-source diode inasmuch as the polarity is appropriate for such action. However, the internal diode must "see" a voltage of about 0.7 volt for this to occur. Such shorting does not occur because the drain-source voltage is *less* than the requisite forward-biasing voltage of the internal diode. Thus, the internal diode does not provide a current path. Our concern with it stems from the fact that a MOSFET with low drain-source resistance must be chosen for use in such a circuit. Otherwise the four internal diodes will form an ordinary diode rectifying bridge in which the overall efficiency will fall short of that attainable from proper participation of the MOSFET portion of the devices.

When the upper-left MOSFET is impressed with "conventionally correct" positive drain voltage during the alternate portion of the ac cycle, its gate is negative. This MOSFET is then turned off. Moreover its internal diode, being reverse-biased, remains out of the picture. It is seen that the unidirectional conduction of this MOSFET exactly simulates the rectifying behavior of an ordinary rectifying diode.

The other MOSFETs operate in the same way; the overall circuit therefore behaves as a bridge-rectifying configuration. Note that two of the MOSFETs must be p-channel devices in order to comply with polarity requirements. The p-channel MOSFETs are conductive in this application when their drain-source voltages are "unconventionally" *positive* and their gates are simultaneously *negative*.

A UNIQUE DRIVE CIRCUIT FOR THE GTO THYRISTOR

It is sometimes interesting to contemplate how one advance in technology becomes even more enhanced through *combinations* with other recent developments. For example, the revived interest in gate turn-off

(GTO) thyristors has led to better performing devices that operate at higher power levels than ever before. And at the same time the frequency capability of these devices has also been extended. It has not been always easy to devise satisfactory drive circuitry, however; the common use of inductors to store and release energy for the turn-off pulse often requires considerable empirical work for optimum results, and sometimes it proves difficult to obtain the flexibility needed for operation under varying duty cycles.

Enter the complementary MOSFET driver. At first only n-channel MOSFET devices were available, but more recently the major manufacturers of MOSFETs have also been marketing p-channel versions. In some cases these are mated with certain n-channel units so that a complementary pair results. In other cases one can choose from various n- and p-channel MOSFETs so that a sufficiently close match exists for many applications. For example, the fact that n-channel devices are inherently faster is of little consequence for most uses; indeed even in the rf region this may not mitigate against the many circuitry and performance features that complementary symmetry offers. The MOSFET complementary-symmetry drive technique shown in Fig. 6-4 provides a simple and straightforward means of controlling the GTO thyristor. It is highly desirable that the negative-going turn-off pulse derive from a low-impedance source—this is readily provided by the MOSFET driver. Also it is generally good practice to maintain the turn-on pulse for approximately the duration the GTO will be on; otherwise the GTO

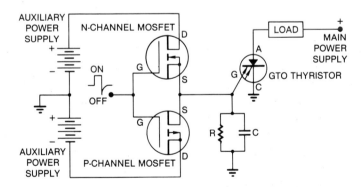

Fig. 6-4. Complementary symmetry MOSFETs as driver for gate turn-off thyristor. The RC network in the gate circuit of the GTO prevents false triggering in noisy environments. Nominal values are several kilohms and several thousand picofarads.

can go out of saturation and develop unnecessarily high dissipation. This is easily provided because the MOSFET pair can be driven from a low-power source.

In actual practice the MOSFETs can be relatively small devices inasmuch as they have high peak-current capability. And the auxiliary power supplies can be put together from a handful of low-cost components. The salient feature of this power control system is that one achieves power-handling ability inherent in thyristors without their nasty commutation problems, especially in switching dc loads. At the same time the high-impedance input is much easier to work with than the low-impedance input of large bipolar transistors. Finally, the peculiar demands of the gate circuit of the GTO are more readily met than by direct-drive schemes. It may not be far-fetched to expect some of the semiconductor firms to market GTO products incorporating this circuitry within a *single package*, either by hybrid or monolithic fabrication.

An Obedient Thyristor—the GTO

Designers of SCR circuits often experience difficulties in inducing or forcing these power devices to turn off once they have been triggered into their on states. The turn-off process, otherwise known as *commutation*, is indeed one compelling reason why transistors are often favored over thyristors. Transistors obediently turn off when their base gate drive signals are removed. But transistors cannot match the power-handling capability of thyristors and are not always cost-competitive for high current and high voltage applications. What has been long desired is a *thyristor* that can be turned off via its gate.

For two decades various companies have introduced such gate turn-off thyristors, or GTOs, only to terminate their production. There have been various reasons for this. Early versions of this device required so much power for the turn-off function that the overall efficiency was compromised. Later types were better performers but did not compete well with other devices, such as Darlingtons and MOSFETs. A more sustained development program has been maintained by overseas semiconductor firms, where there seems to be more grass-roots enthusiasm over the device. The reliability, together with frequency and power capability of recent GTOs, suggests that the device has a definite future both in this country and abroad. It shows promise as a rugged power switch for motor control, inverters, and regulated power supplies. Its ability to operate from a dc source is especially attractive.

Examples of GTO applications will be given using RCA types despite the fact that RCA no longer makes the devices. They were used for several years and proved eminently satisfactory; in many instances they remain available from distributor's stocks. However, other firms make similar devices. Among these are Unitrode, Mullard, and Philips. The latter company makes the BT family of GTOs, among which can be found almost direct substitutes for the RCA devices. Parameter ratings embrace such capabilities as 30-kHz turn-off rate, 15 amperes of controllable load current, and 1,500 volts maximum anode voltage.

The driven inverter circuit of Fig. 6-5 is of relevant interest. The GTO was one of the RCA G5001 series of high-frequency types. A suitable substitute is available from Philips in either the BTV58 or BTW58 family of GTOs. The salient feature of this driven inverter is its 95 percent efficiency in switching a 1,200-watt load at a 20-kHz rate. Implementation of this basic circuit for use in inverters, converters, tv deflection stages, and switch-mode power supplies is obviously worthy of consideration. The voltage level of the negative turn-off pulse varies with different devices; for the GTO used in Fig. 6-5 it is 70 volts. In all

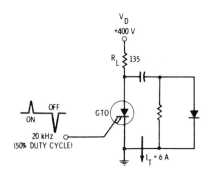

POWER DISSIPATION AND TEMPERATURE FOR 20 kHz
RESISTIVE LOAD SWITCHING WITH I_T = 6 A AND V_D = 400 V

				1200 WATTS			
f	T_J	P_D (ON)	P_D (OFF)	P_{DC}	P_{TOTAL}	T_J - T_C	T_C
(kHz)	(°C)	(W)	(W)	(W)	(W)	(°C)	(°C)
20	100	10	17.4	7.3	34.7	52	48
20	125	10	23.0	7.0	40.0	60	65

Courtesy RCA

Fig. 6-5. Basic GTO-driven inverter circuit. 1,200 watts can be switched with 95% efficiency at 20 kHz. Although a snubber network is used, there is no commutating circuitry.

cases, however, the negative turn-off voltage should derive from a low-impedance source, inasmuch as a high peak turn-off current is characteristic of these thyristors. Snubber networks are generally required to absorb transient energy released by stray or leakage inductance when the thyristor turns off. In Fig. 6-5 the resistor of the the snubber network can be in the vicinity of 100 ohms and the capacitor can be approximately 0.1 microfarad. A fast-recovery type is suggested for the diode in the snubber network.

Inasmuch as no commutation circuitry is associated with the driven inverter of Fig. 6-5, its extension to various control systems is quite straightforward. This is particularly true because pulse-width or duty-cycle modulation is accomplished by simply varying the time interval between the positive turn-on pulse and the negative turn-off pulse applied to the gate. Fig. 6-6 illustrates the basic arrangement of a bridge inverter suitable (with appropriate GTO selection) for both low- and high-frequency applications. Here diodes D1, D2, D3, and D4 are the so-called flywheel of free-wheeling diodes; they provide current paths for release of stored energy in the transformer.

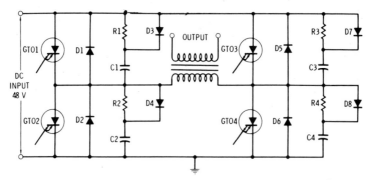

Fig. 6-6. Simplified circuit of GTO bridge inverter. The absence of commutation makes operation straightforward and reliable. Snubbing network resistors R1, R2, R3, and R4 are 100 ohms. Snubbing network capacitors C1, C2, C3, and C4 depend on frequency, but 0.1 microfarad is a good choice for initial evaluation.

Shown in greater circuitry detail is the switched-mode power supply of Fig. 6-7. Here the duty cycle of the incoming square wave controls the output voltage of the supply. The gate-trigger portion of this circuit, which is configured around Q1, not only dispenses with the need for a

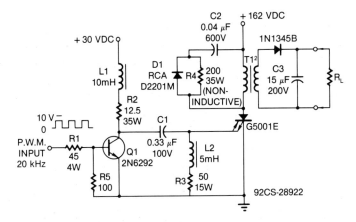

1. Watt values shown are for regulation of 50 to 100 percent loads.
2. T_1 = Core: Siemens E Core Set E55 (M55)
             B66251-A0000-R026
      Air Gap: 0.051 inch
      Primary: 45 turns, 30/36 litz.
      Secondary: 30 turns, 30/36 litz.
      (Sandwiched layers – 1 Pri./2 Sec./ PR1)

Courtesy RCA Solid-State Division

Fig. 6-7. The main portion of a GTO switched-mode power supply (100 volts, 150 watts). A negative voltage source is not required for providing the gate trunoff signals to the GTO SCR.

negative voltage supply but effectively doubles the available 30 volts so that the gate of the GTO receives negative turn-off pulses close to 60 volts in amplitude.

The circuit of Fig. 6-7 utilizes the well-known fly-back principle hitherto widely used with power transistors but seldom with thyristors. The limited use of thyristors has been primarily in tv high-voltage circuits; however, designers of inverters and switching-power supplies operating from dc sources have shunned the thyristor because of the turn-off problems. Yet here we see a power supply making use of the GTO SCR. As suspected, this stems from both this device's high-frequency capability and its turn-off characteristic. The gate-drive transistor, Q1, is able to generate both positive turn-on and negative turn-off pulses for the GTO thyristor. Thus, there is no need for a negative dc source. When Q1 is switched off, the GTO is triggered to its on state by stored energy derived from inductor L1 and also from inductor L2.

When Q1 is switched on, the gate of the GTO receives a negatively polarized turn-off pulse from the charge stored in capacitor C1. The primary winding of isolation transformer, T1, develops sawtooth voltage-pulses in the same manner as the primary of an automotive ignition coil, and these pulses are stepped up and rectified in the secondary circuit. Transformer T1 operates in its linear region—magnetic saturation is not involved in this type of inverter. (The plus 162 volts derive from full-wave bridge rectification and single-capacitor filtering of the 115-volt, 60-Hz line.)

Although this inverter-type power supply is not regulated as shown in Fig. 6-7, the experimenter can readily incorporate a feedback loop for this purpose in the same manner as commonly done in similar supplies using power transistors. Also the circuit of Fig. 6-7 can be "beefed-up" to provide considerably more output power, inasmuch as the GTO is not a flea-power device. To what degree this unique inverter will merit consideration from manufacturers of regulated and high-voltage supplies remains to be seen. The GTO thyristor also offers interesting possibilities for polyphase motor-control applications, cycloconverters, and welding power sources. Perhaps another round of performance upgrading will shatter whatever psychological inertia presently inhibits its more widespread use; it appears to be a safe bet that this is just around the corner.

Another semiconductor firm that has pioneered notable advances in high-power GTO thyristors is the Unitrode Corporation. In particular this company markets such devices that require exceptionally low turn-off voltages and currents—on the order of 5 volts (negative) and several milliamperes. (Although the described RCA units need 50 to 70 volts for turn-off, together with a high peak turn-off current, it is well to keep in mind that there are many interrelated trade-offs in the performance parameters of these devices.)

A NEW HIGH-FREQUENCY SWITCHING DEVICE

A new power device has given a good account of itself in laboratory development and is bound to make a strong inpact in high-frequency switching applications, such as in pulse-width modulated regulators and inverters. For such uses, the bipolar power transistors have featured low voltage drop but high switching losses at high frequencies, say, beyond 75 kHz. On the other hand power MOSFETs have demonstrated their ability to switch efficiently up to about a half-gigahertz insofar as rise and fall times; however, the relatively high on resistance of the MOSFET devices is bothersome when high currents at low and moderate voltages have to be switched. For example, in 5-volt regulated power supplies, it is not always easy to decide whether to live with the rise and fall switching losses of the bipolar transistor or the relatively high on voltage drop developed in the MOSFET device. In any event designers are not partial to compromising their quest for ever-higher switching rates because it leads to space and weight savings through smaller magnetic components and smaller capacitors. Generally such high-frequency technology is also accompanied by reduced manufacturing costs as well.

Because of such factors many designers have secretly wished for the best of both worlds: the fast rise and fall times of MOSFETs and the low on resistance of bipolar transistors. A device undergoing development is indeed a combined unit, comprising a power MOSFET and a power bipolar transistor in parallel; the dual device is shown in Fig. 6-8. The gate and base leads are independently brought out, making the overall device a four-terminal one. The purpose of doing this is to enable sequential timing to be applied to the two transistors in such a way that the MOSFET is first turned on. This provides a quick rise time and absorption of turn-on transients. The bipolar transistor is next turned on and takes over the task of carrying load current once it is fully turned on. It does this by virtue of the fact that its on resistance is considerably lower than that of the MOSFET. The parallel combination is turned off by removing the base drive from the bipolar transistor; after it turns off, the gate signal is removed from the MOSFET, thereby allowing a quick turn-off of load current through the combination. Our combined device provides low switching losses *and* low on losses at high switching rates.

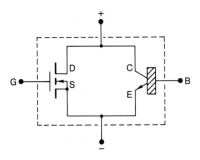

Fig. 6-8. Composite power device produced by paralleling bipolar and FET devices. Although implementable with discrete elements, this combination readily lends itself to monolithic integration as a single module.

The beauty of this forthcoming power device is that the MOSFET and bipolar sections will be monolithically integrated as a *single* module. In the meantime, however, the experimenter can approach the efficient switching performance of this technique by using discrete elements. Discrete devices might even provide some circuitry flexibilities denied in the monolithic combination. In any event the greatest rewards from experimental investigation will probably be forthcoming from the attainment of optimum timing sequence and duration. This will be especially true if duty-cycle modulation is used in order to accomplish regulation in a switching-power supply.

THE CASCODE ARRANGEMENT OF FET DRIVER AND BIPOLAR POWER STAGE

Consider the manifold applications of bipolar power transistors—invariably the common-emitter or common-collector circuit configuration is used. However, the common-base connection is capable of providing a much wider safe operating area than either of these two formats. Moreover, the common-base mode of operation yields a significantly higher switching-rate capability too. Why then has this more natural way of operating a power transistor declined in popularity? The answer is that power transistors have grown greatly in their current-handling capability; this means that the relatively low input impedance of the common-base arrangement is low indeed for high-power transistors. For practical reasons it is difficult and unwieldy to devise a satisfactory driving source for this otherwise-desirable configuration. Enter the power MOSFET.

The arrangement shown in Fig. 6-9 combines a low-voltage, high-current MOSFET with a high-voltage bipolar transistor in the so-called *cascode* connection. In a sense the MOSFET input device functions as an impedance step-down transformer operating all the way down to dc. But no physical transformer could yield the high input impedance of the MOSFET. Although this scheme is implementable with discrete elements, monolithic integration would obviously be desirable. Unfortunately this is not easily accomplished with the devices interconnected in this way. But a hybrid combination of the two types of transistors would be a good compromise inasmuch as the package would be dealt with as a single device featuring all of the aforementioned attributes.

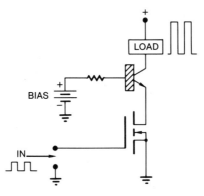

Courtesy International Rectifier Corporation

Fig. 6-9. Basic circuit for cascode operation of bipolar/FET devices. When driven in this manner by the MOSFET stage, the bipolar power transistor readily provides high-speed switching of load and current from a high-voltage dc source. This is achieved with much less voltage drop, $V_{CE(sat)}$, than would be developed across a single power-MOSFET switching circuit.

A natural question with regard to this arrangement is, Why not simply produce a high-voltage power MOSFET? With present technology this would indeed be the usually preferred procedure up to about 500 volts. However, for applications requiring voltage ratings of 800, 1,000, or higher, the voltage drop across suitably rated MOSFETs becomes too high in many instances. All things considered, an evaluation of both circuit performance and manufacturing cost could make the cascode arrangement appear the more favorable approach for many high-voltage switching applications.

The International Rectifier Corporation has trademarked its version of the power MOSFET the HEXFET. Also this firm designates the described cascode format as the BIMOS switch.

THE NEW BREED OF POWER DARLINGTONS

What is so intriguing about the power Darlington is not its basic concept, which is admittedly old hat; rather it is rapid advancement in the technology of its fabrication. Long a candidate for less-demanding applications, it now carries voltage, power, and speed ratings, which makes it useful in circuitries previously monopolized by the discrete power transistor. Its high current gain enables it to be driven from a low-power source, such as logic systems. In former design practice it was common to use a discrete power transistor driven by a discrete small- (or smaller-) signal transistor. Overall the same thing was accomplished as is now attainable via a power Darlington, but with more interconnections, a higher parts count, and greater cost. The power Darlington looms as one of the important workhorses in the power control technology of the future. At present 250-watt units are commonly available, and it is likely that the market pressures of competition will push power-dissipation ratings higher yet.

The $V_{CE(sat)}$ parameter of the power Darlington transistor merits a word or two. Normally the output transistor of a Darlington pair is not amenable to saturation in the sense that a discrete transistor can be driven into this operational mode. This was long cited as one of its severe disadvantages. However, instead of operating from several tens of volts, modern power Darlingtons may operate from sources supplying several hundred volts or higher. Thus, the lack of voltage saturation is not of great importance any more—the circuit efficiency is nearly the same whether the device develops a voltage drop of a fraction of a volt or a drop of several volts. At the same time the nonsaturating feature has become an advantage because it enables much faster switching speed. Thus, the symbol $V_{CE(sat)}$ is somewhat of a misnomer. (Somewhat similarly specification sheets on power MOSFETs allude to "junction" temperature despite the lack of a junction in such a device.)

The four basic Darlington connections are shown in Fig. 6-10. Monolithic integration is most straightforward for circuits (A) and (B). Many of the pnp types depicted in Fig. 6-10B are becoming available as companion units for popular npn Darlingtons of the type shown in Fig. 6-10A. This facilitates the design of complementary-symmetry amplifier stages, which provide push-pull operation without transformers.

Generally the monolithic fabrication of power Darlingtons includes elements in addition to the input and output transistors. A typical example is shown in Fig. 6-11.

Referring to Fig. 6-11, the right-hand diode is there to protect the collector junction of the output transistor. The use of such a diode in circuit connections of discrete devices is quite common—it is seen whenever a power transistor is used to drive a solenoid, motor, or other inductive load subjected to abrupt switching of current. Also this diode

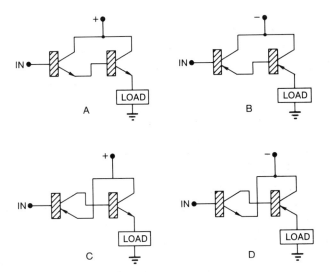

Fig. 6-10. The four basic Darlington circuits. In all of the formats the load-connected transistor is the actual power device, whereas the input transistor is relatively small: (A) npn Darlington circuit; (B) pnp Darlington circuit; (C) simulated pnp Darlington circuit; (D) simulated npn Darlington circuit.

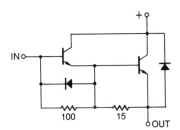

Fig. 6-11. Equivalent circuit of MJ10020 power Darlington transistor. Monolithic fabrication incorporates bias-return resistors and diodes for protection and turn-off speedup.

often serves as a free-wheeling diode to maintain constancy of current flow in an inductive load, such as the inductor of a filter network. It is less evident what the function of the left-hand diode is; this is the so-called speedup diode, but the name does not tell us how it performs its mission. Because of its polarity relative to that of the incoming forward-bias pulses (in a switching-power supply), it appears that this diode's circuit effect is both inert and benign.

Such a superficial observation is partially true—this diode exerts negligible influence as long as only positive-going pulses are impressed at the input terminal. However, this diode will convey negative pulses directly to the base of the output transistor; this is exactly what is needed to "sweep out" stored charge in the output transistor, thereby speeding its fall time. Therefore, in a high-frequency switching supply, negative pulses follow the longer positive pulses. This circuitry technique provides electrical access to the base of the output transistor without the necessity of physically bringing out a fourth terminal.

On the other hand some manufacturers offer power Darlingtons in a four-lead package wherein the base connection to the output transistor is actually brought out to the extra terminal. This provides more design flexibility; its salient feature is that it enables better current sharing when power Darlingtons are paralleled. Fig. 6-12 depicts such a situation.

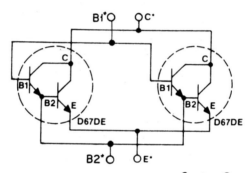

Courtesy General Electric Co.

Fig. 6-12.   Paralleling technique with four-terminal Darlingtons. This scheme promotes equal current sharing under both static and dynamic operating conditions. Note the absence of emitter-ballast resistors, whcih are needed when utilizing ordinary transistors.

OPTO-ISOLATOR DRIVE FOR ANTIPARALLEL SCRs

When the triac became available, one of its heralded features was displacement of the awkward antiparallel configuration of two SCRs. Thus, not only was one discrete device eliminated, but a single gate served as a relatively simple means of control. In contrast antiparallel-connected SCRs required complicated circuitry to generate appropriate control signals for two separate gates. The advantages offered by the triac were self-evident, and designers used this device wherever possible. As expected, triac technology made significant advances, and units with greater power-handling ability appeared on the market. Special 400-Hz units were perfected for aircraft deployment. Notwithstanding the success of the triac, it remains necessary to resort to antiparallel-connected SCRs when dealing with very high-power loads and with frequencies beyond the capabilities of the 400-Hz types.

For example, such uses as welders, ultrasonic equipment, and speed control of integral-horsepower motors tend to fall outside the domain of triac applications; a wide variety of high-power, high-frequency, high-current, and high-voltage SCRs are available to enable direct exploitation of the antiparallel arrangement. Moreover, special opto-isolator modules have been developed to provide the gate-control signals for the SCRs, thereby dispensing with the transformers and associated circuitry previously needed. Not only do such modules provide as good, or better, electrical isolation than transformers, but they also function as zero-voltage switches. This greatly reduces, or eliminates, emi and removes electrical stress from both the SCRs and the load.

An example of such a control module is the Motorola MOC3031 illustrated in Fig. 6-13. The MOC3031 contains zero-crossing circuitry. The basic setup for controlling antiparallel SCRs is shown in Fig. 6-14. The diodes D1 and D2 can be multipurpose types, as the 1N4001. R1 and R2 will usually be in the vicinity of 1 kilohm. (The antiparallel configuration is also referred to as the inverse-parallel or back-to-back connection. As with the triac, full-wave power is delivered to the load.)

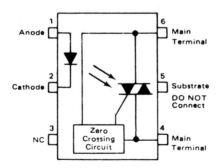

Courtesy Motorola Semiconductor Products, Inc.

Fig. 6-13. Thyristor driver with self-contained zero-voltage-crossing circuitry.

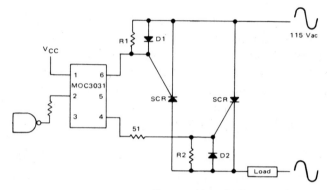

Courtesy Motorola Semiconductor Products, Inc.

Fig. 6-14. Drive scheme for antiparallel SCRs. Opto-isolator module provides zero-voltage crossing and facilitates triggering of antiparallel SCRs in full-wave control arrangement.

The arrangement shown in Fig. 6-14 is particularly well-suited for burst-modulation control, as described in Chapter 1. In this power-control technique a varying number of integral cycles are delivered to the load.

COUPLING OF THE MICROPROCESSOR TO HEAVY AC LOADS

Although a number of dedicated IC interfaces are available for coupling the outputs of a microprocessor to the external world, this transition is not always easily accomplished when one wants the microprocessor to control output devices requiring healthy slugs of ac power from the utility line. For example, it might be desired to energize a large incandescent lamp, operate a solenoid, or run a motor. Such devices, because of their power requirements, are generally controlled by thyristors, such as the triac. This at once introduces certain problems. First, there may be a nasty conflict beween commons or grounds; certainly it would be desirable to isolate the microprocessor and its low-level interface circuitries conductively from the ac power line. In the second place simply turning on and off the ac power supplied to a lamp or to a motor may not be the optimum way of conveying commands from the microprocessor to the loads. This is because the sudden application of voltage to a cold lamp filament tends to shorten the lamp's life. The resultant current inrush can also damage the triac. Somewhat similar considerations apply to a "cold" motor—there is again a current inrush. Finally, the sudden application of line voltage produces emi.

All things considered, what is needed is a transfer device that will boost power levels so that the gates of the triacs can be triggered, one that will provide conductive isolation of the utility power circuit, and one that will enable the load to be impressed with ac power *only* during zero crossings of the 60-Hz voltage wave. A unique device, the Motorola MOC3011 optically coupled triac driver, enables these operating requirements to be met in a straightforward manner and at low cost.

The MOC3011 is essentially a small triac triggered from the infrared radiation of an LED. This thyristor has no electrical gate. It may be considered a "slave" for the triggering of the much larger external triac. The nice thing about the optical actuation is that the device can withstand up to 7.5 kilovolts between input and output. This takes care of all varieties of ground problems, ground loops, etc. The schematic of this simple device is shown in Fig. 6-15. Unlike other opto-isolators this one is made for driving a triac.

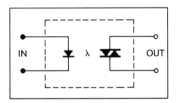

Fig. 6-15.   Schematic representation of the MOC3011 optically coupled triac driver. This is a simpler transfer device than the MOC3031; the triac has no electrical gate and there is no zero-crossing circuitry.

Fig. 6-16 illustrates a real-life application to an M6800 microcomputer system. Although the MOC3011 devices do not themselves establish the zero-crossing constraint, they make this operational mode easy to accomplish via simple circuitry. Thus, the 2N3904 npn transistor works in conjunction with the NAND gates of the 7400 IC and the double line frequency pulses from the full-wave rectifier so that triac firing is inhibited at all times *except* when the 60-Hz voltage wave crosses zero *and* a command is generated by the microprocessor unit (MPU).

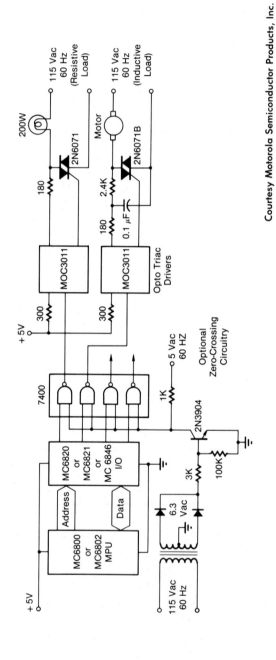

Fig. 6-16. Interfacing a microcomputer system to power-line loads with opto-isolators. The MOC3011 modules are dedicated devices intended for driving large triacs.

NEW FEATURES IN POWER-MOSFET AUDIO

The power-MOSFET audio amplifier shown in Fig. 6-17 takes on special significance when compared with the circuit of Fig. 4-6, which was state of the art several years earlier. Although the earlier circuit remains a good approach to modern high-fidelity practice, the newer circuit is obviously simpler and has a much lower parts count. And even though no resort is made to paralleling of the output devices, the latter circuit has greater power capability. This has been brought about by two developments. First, as expected, power MOSFETs have been growing in power-handling capability. Second, and somewhat surprisingly, families of powerful p-channel MOSFETs have become available. These can be mated to appropriate n-channel devices to form complementary-symmetry circuitry. As with bipolar transistors, complementary symmetry tends to simplify circuit configurations and reduce parts count.

Unlike the quasi-complementary circuit of Fig. 4-6, this power amplifier utilizes *true* complementary symmetry. Both output devices,

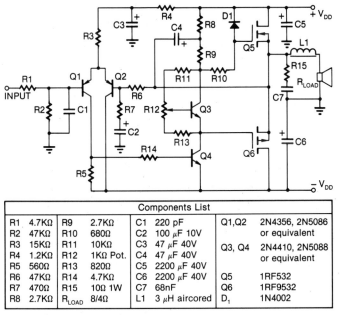

Components List						
R1	4.7KΩ	R9	2.7KΩ	C1	220 pF	Q1,Q2	2N4356, 2N5086
R2	47KΩ	R10	680Ω	C2	100 μF 10V		or equivalent
R3	15KΩ	R11	10KΩ	C3	47 μF 40V	Q3, Q4	2N4410, 2N5088
R4	1.2KΩ	R12	1KΩ Pot.	C4	47 μF 40V		or equivalent
R5	560Ω	R13	820Ω	C5	2200 μF 40V	Q5	1RF532
R6	47KΩ	R14	4.7KΩ	C6	2200 μF 40V	Q6	1RF9532
R7	470Ω	R15	10Ω 1W	C7	68nF	D,	1N4002
R8	2.7KΩ	R_LOAD	8/4Ω	L1	3 μH aircored		

Fig. 6-17. 60-watt complementary-symmetry MOSFET amplifier. High-power MOSFETs have become available in both n-channel and p-channel fabrications.

Q5 and Q6, operate as source-follower or common-collector amplifiers. Because of the high-frequency capability of MOSFETs, the source-follower configuration, with its near-unity voltage gain, is less likely to cause self-oscillation problems than would be the case for common-source circuitry. This also makes sense from an overall circuit viewpoint in that the high input impedance of the MOSFETs facilitates development of the requisite voltage gain in the driver stages.

An interesting aspect of this circuit is the network comprised of C4, R8, and R9. This network delivers "bootstrapped" drive to the gate of output MOSFET Q6, thereby causing symmetrical division of output power between Q5 and Q6. Diode D1 also enters the picture here; this diode functions as a clamp during overload conditions so that symmetry is not grossly impaired by overdrive of Q5. (The other output device, Q6, is protected in a similar manner by the collector-emitter section of driver transistor Q4 and therefore needs no diode in its gate circuit.)

Drive is provided for the complementary-symmetry output stage by Q4, which operates as a Class A voltage amplifier. Part of the collector load of Q4 is comprised of Q3 and its associated circuitry. This enables adjustment of the quiescent current of the output stage by means of potentiometer R12. Optimum Class AB operation of this amplifier occurs when the quiescent current is 100 mA for ± 30 volts delivered from the power supply.

Driver stage, Q4, is in turn driven by the pnp differential input pair, Q1 and Q2. Thus, the overall circuitry of the amplifier is essentially straightforward and is recognizably simpler than other amplifiers with 60-watt capability. In addition to the generalized operating principles just delineated, some specific details follow.

The voltage gain of the amplifier is governed by the value of the feedback network R6, R7, and C2. For practical purposes the gain is established by appropriate selection of R7. For a gain of 100 the value of R7 is 470 ohms; for a gain of 20 the value of R7 is 2.2 kilohms. The frequency response is approximately 15 Hz to 100 kilohertz in either case. However, the 60-watt output can be attained only with a 4-ohm load. About 30 watts can be developed in an 8-ohm load. Resistances R10 and R14 and capacitors C5 and C6 are important in that they prevent instabilities and self-oscillation. This merits special mention because of the power and bandwidth of this amplifier; self-oscillation can result in catastrophic destruction of the MOSFETs. However, if the amplifier is laid out according to good high-frequency practice and is carefully brought into proper operation, there will be no cause for concern.

Although this book has not dealt in constructional details, the PC layout of this amplifier is depicted in Fig. 6-18in order to serve as a guide. MOSFETs have rf capabilities not generally encountered in audio-type bipolar transistors. On the one hand this contributes to flat frequency response without too much reliance on negative feedback; on the other hand one must be inordinately careful to avoid situations leading to self-oscillation.

The thermal situation of this amplifier is relatively easy to satisfy. The drain terminal of the T0220 packaged MOSFETs is electrically connected to the tab. This provides both thermal and electrical benefits in this circuit. Thermally, because no insulator is required between device and heat sink, heat transfer to the sink is optimum. Electrically the possibility of signal feedback is greatly reduced because the heat sink is at ac ground potential. The singular requirement for the heat sinks is that they have less than 1.16 C/W thermal resistance to ambient.

The total harmonic distortion is on the order of 0.15 percent at 60 W into 4 ohms. For an 8-ohm load these numbers are approximately halved. These measurements were conducted at 1 kHz. (The distortion measurements pertain to a voltage gain of 100. As mentioned, operation at a voltage gain of 20 can be had by appropriate selection of R7; the distortion will then be almost proportionately reduced from the cited figures.) The input impedance of the amplifier is 47 kilohm and the bandwidth is approximately 15 Hz to 100 kHz.

The power supply for this amplifier is shown in Fig. 6-19. Despite the topological resemblance to a bridge circuit, what we actually have are two full-wave rectifying circuits operating from a single transformer winding. In this respect the supply is similar to that of Fig. 4-6C. How-

Copper Side

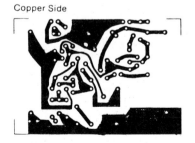

Component Side

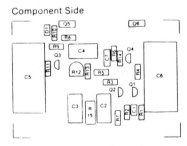

Courtesy International Rectifier Corp.

Fig. 6-18. Printed-circuit board layout for 60-watt amplifier. This is shown for guidance, not as a how-to presentation. The emphasis is on short leads and on placement of parts complying with good high-frequency practice.

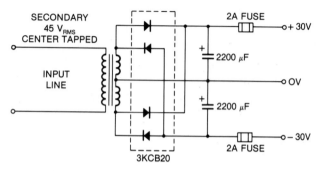

Fig. 6-19. Power supply for complementary-symmetry MOSFET amplifier. A transformer with a rating of 100 VA is a good choice for cool operation; a smaller capacity transformer could be used inasmuch as amplifiers usually are not operated at full power output for appreciable lengths of time.

ever, whereas the supply of Fig. 4-6C is intended for use with two amplifiers, this supply can accommodate only a single amplifier channel. If two-channel operation is contemplated, as in stereo, an additional power supply will be necessary. (Of course, the experimenter can double the current capability of the supply depicted in Fig. 6-19 by using appropriately scaled up components. For example, the value of the filter capacitors should be doubled. But 4-ampere fuses would not serve any useful purpose. Rather 2-ampere fuses should be retained, but they would then be moved directly to the dc inputs of the two amplifiers. Single-supply operation is more cost-effective, but a two-power supply arrangement has definite pluses for the experimenter.)

TUBE POWER LEVELS FROM SOLID-STATE RF DEVICES

Amateur radio operators have long been accustomed to rf power levels, and to performance features provided by the 6,146 beam-power tubes. This is because these tubes, together with their reasonable dc power requirements and their meager drive demands add up to respectable output power at minimal cost. Although the trend toward all solid-state transmitters has been well-publicized, there has been inordinate emphasis on either low power or conversely on sophisticated and expensive kilowatt finals. Somehow the solid-state devices suitable for producing power levels in the 50- to 150-watt range have often been electrically fragile and unreliable and have involved cash outlays not too pleasing to the amateur interested in operating with such modest power. The use of transistors to develop 5 or 10 watts may be economic but would not be likely to compel the junking of the tube rig.

Fig. 6-20 shows the schematic diagram of a broadbanded 2- to 30-MHz linear amplifier capable of 140-watts PEP power. Besides single side-band, it is also useful for 60 watts of AM output power and at least 160 watts of CW power. A driver with 3- or 4-watt capability should

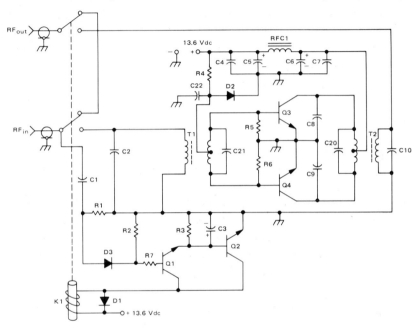

C1	=	33 pF Dipped Mica	R7	=	100 Ω 1/4 W Resistor
C2	=	18 pF Dipped Mica	RFC1	=	9 Ferroxcube Beads on #18 AWG Wire
C3	=	10 μF 35 Vdc for AM operation,	D1	=	1N4001
		100 μF 35 Vdc for SSB operation.	D2	=	1N4997
C4	=	.1 μF Erie	D3	=	1N914
C5	=	10 μF 35 Vdc Electrolytic	Q1, Q2	=	2N4401
C6	=	1 μF Tantalum	Q3, 4	=	MRF454
C7	=	.001 μF Erie Disc	T1, T2	=	16:1 Transformers
C8, 9	=	330 pF Dipped Mica	C20	=	910 pF Dipped Mica
R1	=	100 kΩ 1/4 W Resistor	C21	=	1100 pF Dipped Mica
R2, 3	=	10 kΩ 1/4 W Resistor	C10	=	24 pF Dipped Mica
R4	=	33 Ω 5 W Wire Wound Resistor	C22	=	500 μF 3 Vdc Electrolytic
R5, 6	=	10 Ω 1/2 W Resistor	K1	=	Potter & Brumfield
					KT11A 12 Vdc Relay or Equivalent

Courtesy Motorola Semiconductor Products, Inc.

Fig. 6-20. Schematic diagram of broadband 140-watt PEP linear amplifier. Included are a carrier-operated relay and a bias-compensation circuit.

suffice. There are no variable capacitors or other tuning expedients for resonating, neutralizing, or coupling. (It is conceivable, of course, that the antenna or its feeder line may require some reactance-cancelling adjustments.) The overall circuitry comprises three basic sections: the push-pull amplifier configured around rf power-transistors, Q3 and Q4; the carrier-operated relay circuit involving small-signal transistors Q1 and Q2; and the bias-compensation network, made up of D2, R4, and C22. The RF amplifier itself is not unique; the carrier-operated relay is an optional luxury and could be dispensed with; the bias-compensating scheme is, however, the most important aspect of the amplifier. Let us see why.

An amplifier of this type actually operates in Class AB, for this is the mode of operation resulting in the best blend of efficiency and linearity. Class AB operation requires a quiescent current, i.e., collector current in the absence of rf input from the driver. For this amplifier the quiescent current is in the vicinity of 1 ampere. Without bias compensation it would be difficult to hold the quiescent current sufficiently in the face of the wide temperature excursions of the power transistors. In actual operation this tends to produce nonlinearity and, what is much worse, the destructive phenomenon of thermal runaway. (Higher temperature causes higher quiescent current, which further increases operating temperature, which in turn increments quiescent current yet higher, etc. What was initially a relatively small quiescent current thereby grows regeneratively until the transistor is catastrophically destroyed.)

The important aspect of the bias-compensating circuit is the tight thermal coupling between diode D2 and rf power-transistors Q3 and Q4. This requisite is implemented by mounting D2 in the heat sink by means of a press fit. The basic idea is to have D2 temperature-track the emitter-base sections of transistors Q3 and Q4. To the degree to which this is attained, the quiescent collector currents of Q3 and Q4 will tend to remain constant with respect to temperature.

High-frequency solid-state circuitry, especially when involving high power levels, is more readily transformed into physical hardware when at least minimal guidance is provided for component layout and mounting techniques. To this end Figs. 6-21, 6-22, and 6-23 point the way for good RF and thermal construction practices. Fig. 6-21 depicts the general fabrication technique—a PC board and a heat sink combined as a "chassis." The identification of the various components is shown in Fig. 6-22. And Fig. 6-23 details the constructional relationships of the PC board, the heat sink, and the bias-compensation diode.

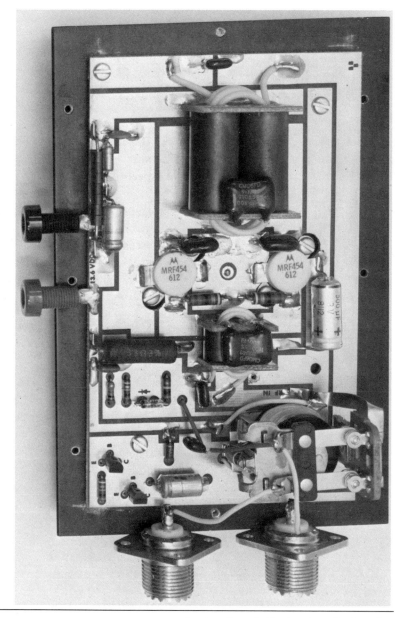

Courtesy Motorola Semiconductor Products, Inc.

Fig. 6-21. Top view of 140-watt PEP linear amplifier.

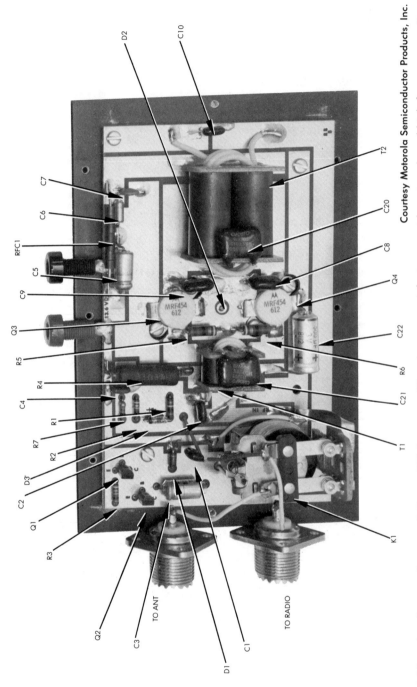

Fig. 6-22. Identification of parts in 140-watt PEP linear amplifier. Arrangement features neat layout and good high-frequency practice.

Courtesy Motorola Semiconductor Products, Inc.

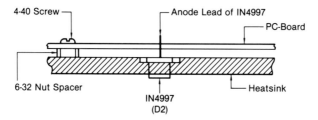

Fig. 6-23. Construction details showing the PC board, heat sink, and bias-compensation diode.

The important performance parameters of this amplifier are shown in Figs. 6-24 and 6-25. These measurements were made at 30 MHz because this represents worst-case operation. Although the amplifier is rated at 140-watts PEP, actual gain saturation does not set in until approximately 210 watts are developed in the load. This contributes to the electrical ruggedness of the amplifier and allows for maladjustments and experimental techniques. The 13.6-volt dc requirement makes this amplifier particularly useful for mobile operation. Because of its power capability, its use for CB operation is illegal. Also broadbanded amplifiers of this type generally require a low-pass filter in the antenna feeder line. A different filter must be used for each band, although a common filter often suffices for both 10 and 15 meters. *Within* each HF

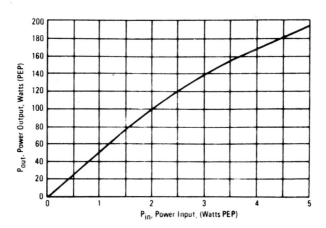

Fig. 6-24. Output versus input power relationship of 140-watt PEP linear amplifier at 30 MHz. It can be seen that reasonable linearity prevails to the design-stipulated output level of 140 watts PEP.

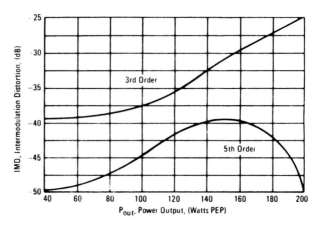

Fig. 6-25.   Intermodulation distortion of 140-watt PEP linear amplifier at 30 MHz.

band the amplifier will operate equally well for any frequency set by the VFO or crystal oscillator.

A final caution is in order. No transistor is as forgiving as an electron tube; it therefore behooves the constructor and operator to exert a little extra effort in getting things right the first time.

GROWN-UP MOSFETs—THE NEW WAY TO HIGHER RF POWER

One of the reasons that bipolar power transistors have made such spectacular gains in rf applications is that their king-of-the-mountain status has been threatened by the relatively new power MOSFET devices. Indeed it appears that bipolars and MOSFETs are embarked on a protracted period of competitive development. This, of course, bodes well for progress in both devices. In all probability one device will not render the other obsolescent; rather each will compel the attention of designers for certain performance parameters that they wish to optimize.

It is accordingly relevant to investigate an approximate MOSFET equivalent of the bipolar linear amplifier just dealt with. The rf power applications of the MOSFET discussed in Chapter 4 remain valid, but considerably extended performance at higher power levels is available from power MOSFETs. Many of these devices are *specifically* intended for rf service by virtue of packaging, internal impedance-match provisions, and other features. Improved thermal design techniques, together with larger dies, have resulted in MOSFET devices that provide a real challenge to their bipolar counterparts. For example, the MOSFET linear amplifier shown in Fig. 6-26 permits operation at a power level of 250 watts PEP. This makes it very suitable for single-sideband work. It may come as a surprise to see that this MOSFET amplifier develops even greater power than the bipolar circuit of Fig. 6-20. No attempt is made, however, to proclaim either type of transistor as having inherently higher power capabilities. Much depends on other factors, such as frequency, linearity, operating voltage and current, heat removal, and cost. We may safely conclude that the power MOSFET is a viable rf device, capable of processing rf at respectable power levels.

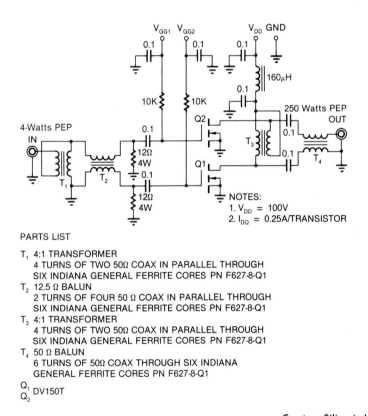

PARTS LIST

T_1 4:1 TRANSFORMER
    4 TURNS OF TWO 50Ω COAX IN PARALLEL THROUGH
    SIX INDIANA GENERAL FERRITE CORES PN F627-8-Q1
T_2 12.5 Ω BALUN
    2 TURNS OF FOUR 50 Ω COAX IN PARALLEL THROUGH
    SIX INDIANA GENERAL FERRITE CORES PN F627-8-Q1
T_3 4:1 TRANSFORMER
    4 TURNS OF TWO 50Ω COAX IN PARALLEL THROUGH
    SIX INDIANA GENERAL FERRITE CORES PN F627-8-Q1
T_4 50 Ω BALUN
    6 TURNS OF 50Ω COAX THROUGH SIX INDIANA
    GENERAL FERRITE CORES PN F627-8-Q1
Q_1
Q_2 DV150T

Courtesy Siliconix Inc.

Fig. 6-26. Broadband 2 to 30 MHz linear amplifier using rf-type power MOSFETs. Quarter-kilowatt PEP can be delivered to a 50-ohm antenna feeder line.

As with the discussed bipolar amplifier, the MOSFET circuit of Fig. 6-26 is broadbanded from 2 to 30 MHz. It has advantages and disadvantages with respect to the bipolar circuit. MOSFETs tend to be immune to thermal runaway; this property often proves rewarding in practical situations where a high vswr may be encountered, say, from an open or short in the antenna system. As a corollary of this feature, no bias-compensation circuitry is required. More Siliconix DV150T MOSFETs can probably be paralleled in order to boost power output to even greater levels. On the other hand the 100 volts required from the dc operating source does not make this arrangement readily adaptable for vehicular use. (The gate-voltage source can be similar to that shown in Fig. 4-21 for the low-power MOSFET amplifier. Or, for better optimization of amplifier linearity, two separate, but variable, sources of gate voltages can be used.)

As with the discussed bipolar amplifier, two operating stipulations apply. This amplifier, because of its power, is illegal for use in the citizen's band. And a low-pass filter will generally be necessary in the antenna feeder line.

BETTER PUSH-PULL OPERATION FOR SINGLE-PACKAGE MOSFET PAIR

Another interesting development in rf MOSFET devices is the push-pull pair designated by Siliconix as DV28120V. There is more than one would infer from casual inspection here, for it is certainly no technological breakthrough to incorporate more than one device within a single package. It happens, however, in high-frequency rf circuits that stray inductance from connecting leads makes it practically difficult to ground the emitter or source terminals of rf transistors properly. Also stray capacitance tends to agitate this situation, giving rise to parasitic oscillation or to erratic frequency response. In push-pull configurations the two transistors often do not operate so as to balance out, or attenuate greatly, second-harmonic distortion because of this rf grounding problem. In other words one falls short of true push-pull operation even if the transistors are closely matched in characteristics.

These malperformances can be largely remedied by the two-in-one packaging technique shown in Fig. 6-27. Indeed, if it were not for the necessity of completing the dc circuit, it would be unnecessary to ground the emitter or source leads physically when both devices are

Fig. 6-27. RF package of the DV28120V push-pull MOSFET device. This scheme overcomes problems from the inductance of the source grounding lead.

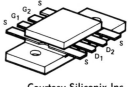

Courtesy Siliconix Inc.

packaged together. This is because this packaging configuration facilitates the tendency of the push-pull circuit to establish a "phantom" ground of zero rf potential at the junction of the emitters or sources of the transistors. This phantom ground is much better than can be accomplished via hard wiring. Therefore, the ground connection serves the dc circuit and is not burdened with the task of providing a tiny impedance as is the case in single-ended amplifiers. As if this were not enough, the DV28120V is a giant in its own right with a 25-degree case-temperature-rated dissipation of 240 watts. And it has 120-watt output capability at 175 MHz. A typical amplifier chain utilizing this dual device is shown in Fig. 6-28.

There is yet another compelling feature to the DV28120V. In rf circuits the input and output impedances of the power device can become

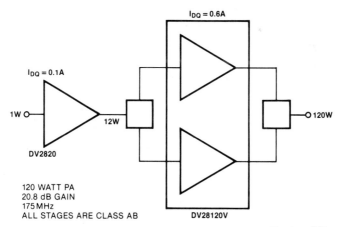

Courtesy Siliconix Inc.

Fig. 6-28. Typical amplifier line-up using the DV28120V push-pull MOSFET device. The rf balance between the two push-pull elements is better than can be readily achieved with discrete power devices.

awkwardly low at high frequencies. Although MOSFETs are superior to bipolar transistors in this respect, another worthwhile improvement is realized by using push-pull, rather than single-ended, amplifiers. The improvement in both input and output circuits is not twice but *fourfold*. Other things being equal, it is physically and practically easier to deal with an input impedance of, say, 4 ohms than of 1 ohm. (The "near-infinite" input impedance of MOSFETs pertains only to dc and low frequencies—at high frequencies the input capacitance makes itself manifest. Nonetheless the rf input impedance of MOSFETs remains appreciably higher than that of bipolar transistors.)

ADDITIONAL INFORMATION ON THE RF AMPLIFIERS

Several variables tend to make the selection of a heat sink for rf amplifiers of the types described both an art and a science. The calculation of allowable temperatures, if made on the basis of the PEP, will result in a heat sink more massive and expensive than necessary under actual operating conditions. This is because of the high ratio of peak to average power in the human voice. Rather miniscule heat sinks can suffice because of this situation. Yet it is not wise to rely on skimpy heat-removal capability. There are times when performance-degrading or destructive temperature rises can easily develop. For example, during two-tone evaluation of single-sideband performance, average power, together with power dissipation, will be considerably greater than that prevailing during ordinary speech. Also, during adjustment of the antenna loading, experimentation with feeder-line filters, optimization of drive conditions, and other tune-up and measurement procedures, the operating conditions of the transistors can involve high dissipation with runaway or destructive temperature rises taking place in surprisingly short time.

In actual practice it is found that a 250-watt PEP linear amplifier can have a heat sink with a specified thermal resistance to ambient in the vicinity of 0.4 degree C/W or less. Proportionate scaling factors can be used for other power output ratings. During all tests, measurements and evaluations before actual voice operation, a fan or blower should be directed at the heat sink. Aside from these obvious precautions to prevent catastrophic destruction of costly semiconductors, better all-around performance invariably attains when the temperature rise of the transistors is minimal. With this in mind it will often be necessary to arrive experimentally at the best heat sink, particularly if the objective is to employ one as small as possible.

WARNING

A widely used ceramic in large power transistors is beryllium oxide, an excellent heat conductor. However, this material should not be crushed, ground, or abraded because it is hazardous to inhale or ingest it. Prescribed disposal is by burial.

A suitable antenna feeder-line filter is shown in Fig. 6-29. This double-pi network is of the constant K, Chebyshev variety, and is relatively easy to design and implement if certain basic precautions are heeded. Make sure that there is no inductive coupling between inductors L1 and L2. This can be achieved by shielded compartments, by arranging these inductors at 90 degrees with respect to one another, or by using toroidal inductors. By the same token the input and output of the filter should be physically arranged so that there is negligible capacitive or radiative coupling enabling harmonic energy to bypass the filter proper. This is readly taken care of by the usual coaxial cable hardware and via the use of a closed metal housing for the filter. (If metal walls are too close to cylindrical coils, both detuning and degradation of attenuation will occur as a consequence of the adverse effect on the Q of the coils.)

"Cutoff" frequencies are used in calculation of the filter elements. These cutoff frequencies, for practical reasons, have to be somewhat *higher* than the high-frequency edge of the respective amateur bands. Suitable cutoff frequencies are 35 MHz, 25 MHz, 15 MHz, 10 MHz, 5.5 MHz, and 2.5 MHz. With this in mind the element values of the filter

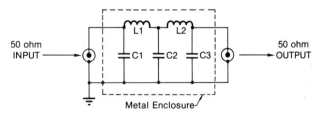

Fig. 6-29. Antenna feeder-line filter for use with broadband amplifiers. 5% dipped mica capacitors may be used. Air-core inductors are suitable; however, below 10 MHz, ferrite toroids are often advantageous. An example of such core material is micrometals grade 6.

are easily calculated using the following information: L1, L2, C1, and C3 should have a reactance, X, of 50 ohms; C2 should have a reactance of 25 ohms. These calculations are made by means of the two equations: $L = X/2\pi f_c$ for inductors and $C = 1/2\pi f_c(X)$ for capacitors.

For example, the filter elements for the 21-MHz band would be calculated using the cutoff frequency, f_c, of 25 MHz. Then

$$L1 = L2 = 50/(2\pi)(25 \times 10^6)$$

$$= \frac{50 \times 10^{-6}}{(2\pi)(25)}$$

$$= 0.318 \times 10^{-6} \text{ H, or } 0.318 \text{ microhenry}$$

And

$$C1 = C3 = \frac{1}{(2\pi)(25 \times 10^6)(50)}$$

$$= 127 \text{ picofarads}$$

Finally,

$$C2 = \frac{127}{2} = 63.5 \text{ picofarads}$$

As a practical expedient short-circuit C2, then check the resonance frequencies of L1-C1 and of L2-C3. If these resonance frequencies are not f_c, modify L1 and L2 accordingly. In this example, $f_c = 25$ MHz. A grid-dip meter is often used for such measurements, in which case best results require as loose coupling as possible to the inductors undergoing measurement.

For a wider and deeper view of rf techniques and applications, the reader should consult the author's book, *Solid-State High-Frequency Power* (Reston Publishing Company).

A WORKHORSE SWITCHING REGULATOR USING A PNP DARLINGTON POWER TRANSISTOR

Although no longer predicated on fact, some designers instinctively shy away from the consideration of pnp or Darlington transistors for applications where a power device must have electrical ruggedness. It has already been pointed out in this chapter that modern Darlingtons not only display respectable power capability but have already established a noteworthy track record in reliability. And pnp bipolar devices, despite their absenteeism from the earlier ranks of silicon power devices, *now* constitute a well-deserved option to npn implementations. Despite the double apprehension some may vestigially feel about pnp Darlington devices, enough time has passed to recognize the viability of these devices as fully acceptable components of solid-state power electronics.

The workhorse switching power supply shown in Fig. 6-30 operates from a nominal 28-volt dc source and can supply up to 10 amperes over an adjustable range of 4-16 volts. As inferred, the switching device is a pnp Darlington power transistor. The circuit of this regulated supply is basically simple, being of the voltage step-down, self-oscillatory type. The basic configuration for such an elemental switching regulator is shown in Fig. 6-31. The difference between Figs. 6-30 and 6-31 comes about through the addition of several refinements, which will be discussed. But first let us investigate the basic operating principle of Fig. 6-31.

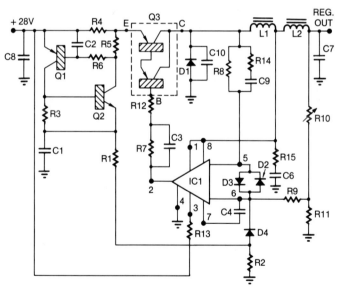

Courtesy RCA

Fig. 6-30. Schematic diagram of self-oscillating regulator with pnp power Darlington.

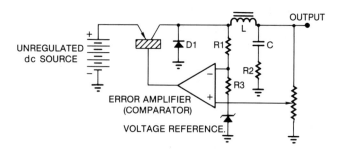

Fig. 6-31. A bare-bones circuit of a self-oscillatory switching regulator. This elemental arrangement helps explain the operation of the more complex circuit of Fig. 6-30.

The Operation of the Basic Regulator Circuit of Fig. 6-31

Despite the configurational simplicity of Fig. 6-31, a characteristic of its operation is that several things take place at one time, and this often leads to awkward design contradictions. Those familiar with linear series-pass regulators will note considerable resemblance between basic linear and switching regulators. For example, the noninverting input of the error amplifier in Fig. 6-31 samples a divided-down portion of the output voltage just as in a linear regulator circuit. Also the reference voltage is applied to the inverting input in an essentially similar way in both types of regulators. In both circuits the comparison between the sampled voltage and the reference voltage affects the error amplifier so as to cause the output voltage to maintain itself substantially constant despite variations in load current or in unregulated input voltage. Nonetheless the manner in which this servo action comes about is different in the two regulator types.

An important difference in operating mode has to do with the operational amplifier, sometimes designated as *error amplifier* but more accurately described as a *voltage comparator* in the switching regulator. Instead of responding to its input signals linearly, the comparator operates in a *bistable* mode—it either produces an output that is all the way positive or all the way negative.

In order for an op-amp to be made to function as a bistable voltage comparator, it must be caused to have a certain amount of hysteresis; that is, it must have a dead zone. This is brought about by resistor R1 in Fig. 6-31. However, R1 simultaneously serves two other functions; it provides the positive feedback path, which makes the entire regulator self-oscillatory, and it provides current to the Zener diode reference. As might be suspected, the value of R3 must be compromised to satisfy these three requirements reasonably. Priority tends to be given to the setting of the hysteresis or dead voltage zone, however. Why this is so will be seen shortly. Resistor R3 is not desirable inasmuch as it degrades the zener reference voltage. It is necessary in order to prevent the zener diode from shorting out the ac voltage at the inverting input terminal $(-)$.

So far we have dealt with a negative feedback path, which serves to correct changes in the output voltage of the regulator, and a positive feedback path, which promotes oscillation. An important aspect of the

---

TABLE 6-1.   Parts List for Self-Oscillating Regulator with PNP Power Darlington

R1	4.7 K	C1	0.1 μF	Q1	2N4036 pnp
R2	1.0 K	C2	0.1 μF	Q2	2N2102 npn
R3	1.0 K	C3	0.1 μF	Q3	RCA 8350B PNP-Darlington.
R4	0.05 ohm, 10 W	C4	0.001 μF	IC1	RCA CA3085
R5	10 ohms	C5	0.1 μF	D1	D2412
R6	100 ohms	C6	50 μF low ESR electro.	D2	1N5392
R7	150 ohms		mylar or polypropylene	D3	1N5392
R8	220 K		can be used. 150 V	D4	1N904
R9	1.0 K	C7	50 μF, 150 V electrolytic.		
R10	10 K	C8	1000 μF, 150 V electrolytic.		
R11	1.0 K	C9	25 pF mica or mylar		
R12	150 ohms	C10	0.02 μF		
R13	300 ohms	C11	0.1 μF		
R14	12.0 K				
R15	Select by experiment. If too low, operation may cease or be unstable; if too high, excessive output ripple will be generated.				
L1	36 turns no. 16 bi-filar on EI75 square stack, grain-oriented silicon steel				
L2	17 turns no. 14 bi-filar on EI75 square stack, grain-oriented silicon steel				

All resistors, except where indicated, are 1/2 W, 5 percent composition.
All capacitors, except where indicated, are ceramic or mylar.

Courtesy RCA

oscillatory behavior is that it involves the triggering of the comparator by the *superimposed ripple on the sampled dc output*. Thus, the operation of this type of switching regulator is dependent on a certain amount of ripple in the output. (If the output were "pure" dc, the regulator would not operate.) It happens that the sampled ripple must have a peak-to-peak amplitude equal to or greater than the voltage hysteresis of the comparator. Again we are faced with design interdependencies. An output capacitor, C, that is too "good" might prevent operation because of its ripple-suppressing action.

This brings us to R2. This resistor may or may not appear in circuit diagrams as a physical component because electrolytic capacitors often have internal resistances that provide the "spoiling" action. Unfortunately they often have too much effective series resistance, and this causes the regulator output to have more ripple than necessary. Inductor L converts the output current from the switching transistor into a triangular waveform particularly suitable for triggering the comparator. L also isolates the switching transistor from damaging peak currents. Thus, the design of such a regulator does not make use of resonance or "filter" formulas to set the values of the output LC circuit.

D1, known as a *commutating*, *catch*, or *free-wheeling* diode, enables the energy stored in the inductor, L, to be used advantageously. Thus, during off intervals of Q1, this diode discharges L through the load. In other words a near-constant load current flows despite the chopping action of switching transistor Q1.

Inasmuch as this type of regulator is so dependent on its output ripple voltage, it should not surprise us that its operation can be sensitive to load, and especially to capacitive loads. With this in mind our investigation of the actual regulator of Fig. 6-30 should prove relevant and straightforward.

Operation of the Practical Regulator

Let us turn our attention to the practical regulator circuit of Fig. 6-30. The essential differences between the basic regulator of Fig. 6-31 and the practical circuit of Fig. 6-30 are
- The practical circuit does not use a discrete zener diode for voltage reference.
- The practical circuit features a two-section output filter.
- The practical circuit includes a current-limiting scheme (and associated circuitry).

With regard to the zener diode, let it first be realized that diode D4 is not a zener type, nor is it operated in its zener-breakdown region. Indeed a first look at the position of D4 can be deceptive, for it is not even connected properly to function as a voltage reference source. Actually D4 is associated with the current-limiting function and can be considered to be out of the circuit except when the regulator operates in its current-limiting mode. The zener-diode voltage reference is, however, contained *within* the monolithic circuitry of the error-amplifier module, IC1.

The two-section output filter is a noteworthy refinement over the single-section filter of the basic regulator circuit. With a single-section output filter a self-oscillating regulator tends to be sensitive to the load and may exhibit erratic behavior under some load conditions. This is because this type of regulator is triggered through its on and off switching levels by the ripple superimposed on the dc output voltage. A load, such as one containing a large capacitive component, that affects the ripple also can play havoc with the switching performance of the regulator. In the circuit of Fig. 6-30 the ripple is determined by L1, R15, and C6 but is effectively isolated from the load by the second filter section, L2, C7.

When excessive load current is drawn by the load (over 11 amperes) the voltage drop across resistor R4 places sensing transistor Q1 in its conductive state, and this in turn supplies forward-conductive bias to transistor Q2. As long as Q2 is on, the hold-off action of diode D4 is overcome, and the resultant positive voltage applied to pin 6 (non-invert) of the error amplifier prevents the regulator from supplying any greater load current.

The two diodes, D2 and D3, protect the input of the error-amplifier–comparator from damaging transients. The CA3085 is actually a complete linear regulator; it is RCA's version of the widely used 723 IC. In this application the CA3085 functions as the error-amplifier–comparator with a voltage hysteresis of about 125 millivolts. (Hysteresis is governed by R8, which also provides the positive feedback to make the regulator oscillate.) Inasmuch as the CA3085 has its self-contained voltage reference, no zener diode appears in the circuit of Fig. 6-30.

A possible point of confusion may arise with regard to the identification of the invert and noninvert terminals of the CA3085. This is because the manufacturer has provided two alternate outputs, one being inverted with respect to the other. However, by following the circuit of Fig. 6-30 and paying heed to the pin numbers, the proper connections will be made during construction. These connections will be analogous to those shown in the basic regulator of Fig. 6-31.

It is characteristic of this type of regulator that the switching rate varies widely, depending on voltage and load conditions. If adjusted for an output of 5 volts, the frequency range will be from approximately 12 kHz at no load to 23 kHz at the full-load output of 10 amperes. If the regulator is adjusted for a 12-volt output, the switching frequency will range from about 18 kHz to 29 kHz over the same load excursion. (These statements assume a nominal dc input voltage of 28 volts.)

The highest operating efficiency of the regulator will obtain in the 2- to 8-ampere load region and will be 80 percent or slightly higher. The efficiency throughout the same load region when the regulator is delivering an output of 5 volts will be 70 percent or slightly better. Interestingly 90 percent efficiency can be approached for an output of 15 volts and 6 amperes, again with 28 volts supplied from the unregulated source. Such efficiency from a relatively simple circuit is bound to alter any preconceived notions about pnp technology, as well as the conductive and switching losses in Darlington power transistors.

SIMPLE SPEED CONTROLLER FOR
INTEGRAL HORSEPOWER DC MOTORS

Ever since solid-state power devices have been available, the control of electric motors has been an intriguing application area. Many advantages have been realized, such as the elimination of power-dissipating rheostats, the deliberate tailoring of motor performance characteristics with less dependence on the "natural" parameters of the motor, and ease of remote control. Much of this progress has been with fractional horsepower motors. Although integral horsepower motors have also been electronically controlled via solid-state power devices, the systems for accomplishing this have tended to be somewhat awkward, complex, and quite expensive. There have also been problems with reliability and electrical ruggedness because of inordinate resort to paralleling and because of the high parts count associated with too many discrete devices.

Recent progress has made available giant transistors and integrated-circuit control modules. By combining these two improvements, it is possible to implement simple control systems for dc integral horsepower motors. And such controllers entirely dispense with the commutation problems almost inherent in SCR control of dc power. Lambda Semiconductors specializes in components admirably suited to the new design approaches for control of dc integral horsepower motors, such as those used in electric vehicles. For example, this company makes a monolithically integrated switching regulator with a 3-ampere output capability. This component can also be used as a pulse-width modulator and driver for a current-boosting stage. Another novel device is a 30-ampere, 100-volt switching-type Darlington with a rated power dissipation of 240 watts. By combining these high-performance products, a simple controller for dc integral motors can be readily implemented.

Such a straightforward control scheme is depicted in Fig. 6-32. One or two additional Darlingtons of the same type could conceivably be paralleled for even greater power-handling capability. On the one hand the control IC develops adequate drive power; on the other hand these power Darlingtons have high current-gain factors. By the same token one could substitute a 50-ampere Darlington recently introduced by this firm, the PMD18D100.

Although this circuit was developed for use with ferrite-ceramic permanent-magnet motors, a little experimentation should result in good

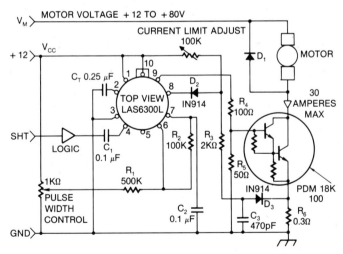

Courtesy Lambda Semiconductors

Fig. 6-32. Integral horsepower dc-motor speed control. Simple scheme extends efficient speed control from fractional to integral horsepower dc motors.

performance with series motors or shunt-field motors. The 25-kHz switching rate advantageously utilizes the inductance of any of these dc motors so that smooth and efficient operation obtains at both slow and fast speeds. (Modern PM motors are often capable of developing comparable starting torque to series motors and may prove more practical to reverse—just transpose the two connections. With a series motor, reversal is accomplished by reversing the connections of *either* the armature or the series-field winding but *not* of both.)

As might well be imagined, the use and control of integral horsepower motors can involve quite abusive electrical conditions; it is imperative that a solid-state system implemented for such control should incorporate means for protecting both electronics and motor from potentially damaging stresses, such as overcurrent and excessive temperature rise. Previously this required complex, and often unreliable, auxiliary circuits and subsystems. With the LAS6300, however, much of this vital task is accomplished internally and it is only necessary to feed in the appropriate signal-level warning voltages. The essential nature of this self-contained protection can be gleaned from inspection of the LAS6300L block diagram shown in Fig. 6-33.

In Fig. 6-33 it will be seen that thermal shutdown and remote control are achieved by inhibiting the switching pulses. This is done by suppressing the output from the comparator. Thermal shutdown is automatic, but remote control via the SHT pin is either a manually actuated function or is brought about from a signal from a transducer or from a feedback provision. In any event a momentary logic 0 turns the LAS6300L on and a voltage above 0.75 volt turns it off. (The experimenter should be aware that thermal shutdown will be defeated if the dc resistance from the SHT pin to ground is less than 5,000 ohms.)

On the other hand, if excessive current is demanded from the LAS6300L, it will automatically go into its current-limiting mode. In this mode of operation complete shutdown of the switching pulses is not produced. Rather the switching frequency is lowered; this results in a lower duty cycle, and therefore in reduced output current. This takes place automatically inasmuch as the sensing resistance is contained within the IC itself. A similar current limiting occurs when the motor current tends to become excessive; this is brought about by an actuating voltage applied to the CLS pin. It will be seen from Fig. 6-32 that this actuating voltage derives from the 0.03-ohm resistance in the Darlington motor circuit. The 100,000-ohm variable resistance provides adjustment of the allowable motor current before limiting sets in. This can prove

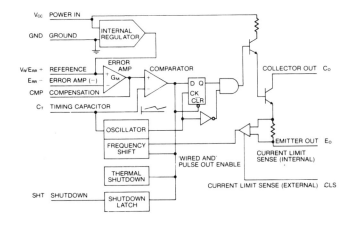

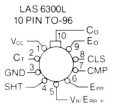

LAS 6300L
10 PIN TO-96

Fig. 6-33. Block diagram and pinout of the LAS6300L. When associated with an external power device, the internal protective circuitry of this IC helps make practical the control of integral horsepower dc motors.

useful in an electric vehicle where there are occasions demanding unusually high motor torque. Of course, such an adjustment must be implemented with discretion in order not to become self-defeating.

Another important protective function is provided by the freewheeling diode, D1. Although the primary function of this diode is to maintain current flow through the motor during off intervals of the switching process, this diode also protects the Darlington switching transistor from inductive voltage transients. A fast-recovery diode with a voltage and current rating at least equal to that of the Darlington transistor should be used—ordinary rectifier diodes are not suitable. Hitherto Schottky diodes, being extremely fast devices, have served this function for systems in which the motor voltage did not exceed 40 or 50 volts. Ordinary Schottky diodes exhibit too high reverse-current leakage at higher voltages. There have been rumors that both domestic and overseas semiconductor firms hope to market a 100-volt Schottky diode in the near future. At present gold-doped diffused-junction diodes, known as fast-recovery types, are quite suitable. Specifically an appropriate selection for D1 would be the Unitrode UES602, which is rated at 100 volts and 30 amperes and has a reverse-recovery time on the order of 50 nanoseconds.

The +12 volts needed for the LAS6300L IC can be derived from a simple auxiliary dc source with a 3.5-amp current rating; in a vehicle this supply can be the first 12-volt battery above ground in the series string of vehicle batteries.

THE FUTURE SOLID-STATE POWER SCENARIO

There are always so many prognosticators that the law of probability suggests a goodly measure of on-target shots. This appears to be true whether the seer is dealing with the weather, the stock market, or horse races. In electronic technology one can hedge chances of true prophecy by not going too far out on the limb; that is, it is a pretty safe bet that power capabilities will be increased, frequency limits will be pushed back, and more sophisticated ICs will be developed. The big obstacle in the prediction art is the abrupt emergence of unanticipated technological breakthroughs. Who, for instance, was preparing us for the advent of the semiconductor transistor when the vacuum-tube reigned as the "obvious" electronic control device? And in the more recent past would one have been a wise gambler to wager against the likelihood of flea-power MOSFET devices suddenly maturing with power-handling capability sufficient to challenge the entrenched bipolar power transistor? For all we know, an entirely new class of materials, say organic substances or armorphous compounds, may one day supersede our silicon, germanium, gallium arsenide, indium phosphide, etc.

To be a predictor and a face-saver obviously involves contradictory techniques of crystal-gazing. Perhaps the best hedge in this endeavor is merely to attempt a near-term extrapolation of presently observed trends. And that, rather than a resort to science fiction fantasies, is exactly what shall be attempted.

Power MOSFETs versus Bipolar Power Transistors

The power MOSFET will continue to extend its domain in the field of power control. Indeed it already appears that the prime reason that these devices haven't made an even greater impact on solid-state power electronics is because of their provocation of improvements in bipolar devices; in order to meet the new challenge, the makers of bipolar devices have come up with giant monolithic Darlington transistors which have been pressed into service in applications where they hitherto had no business, such as in motor control and in high-frequency (20-kHz) inverters and switching regulators. And rf bipolar transistors have had to undergo remarkable upgrading to compete with the MOSFET. All things considered, the large overlap of applications where either device may fit appears to ensure against one device rendering the other obsolete. Rather both devices are likely to evolve continually and competitively with an endless sequence of improvements.

Thus, the successes of bipolar transistors in the 1,000- to 1,500-volt region provide a strong incentive for the development of power MOSFETs for service in the 1,000 volts-plus region. And although 500- or 750-ampere MOSFETs do not appear to be in the near-term mill, their current capability comparison with bipolar can prove somewhat deceptive. For in making such a comparison one must take into account that MOSFETs generally have superior peak-current capabilities; that MOSFETs have more practical safe operating areas; that they have a positive temperature coefficient of resistance, which mitigates against thermal runaway; and that these devices are conveniently paralleled without need for energy-wasting ballast resistances. Therefore, the current handling ability of future MOSFET systems may be largely a function of informative specmanship. On the other hand the manufacturer may provide greater current ratings by packaging two or several MOSFETs in a package with the paralleling connections done internally.

Another source of deception in comparing current capabilities of bipolars and MOSFETs has to do with the effect on the gain of the respective devices. With heavy currents the current gain of bipolar power transistors tends to decrease and may actually impair the usefulness of the device. But with power MOSFETs the gain (or, more apropos, the transconductance) increases with increasing drain current. Finally, it should be pointed out that heavy current in bipolar transistors tends to pull them out of saturation; this causes overheating in switching circuits and can be destructive.

All things considered, it appears that power MOSFETs, power transistors, and power Darlingtons will compete in the arena of moderate power control below 100 kHz. Indeed these three solid-state power devices will usurp applications previously delegated to thyristors. At higher power levels, however, thyristor control appears destined to reign supreme. Whenever one enjoys the luxury of considering any of the four mentioned power devices, either the transistor, the Darlington, or the MOSFET will win over thyristor control, which tends to be plagued with commutation problems and which generates inordinately high rfi and emi.

In rf applications, bipolars and MOSFETs are running neck and neck with the exception that MOSFET superiority mounts up as we go into vhf and uhf regions. Being inherently a higher frequency device (no storage problems from minority carriers), the MOSFET seems likely to provide more bang for the buck at frequencies above about 30 MHz. Interestingly, because some of the giant power MOSFETs have not yet been offered in appropriate rf packages, the potential of this device for processing high power at high frequencies cannot be considered fully exploited yet. Accordingly it can be anticipated that the makers of bipolar transistors or tubes for rf service will be allowed to rest on their laurels.

As mentioned in this chapter, we are on the threshold of seeing monolithically integrated bipolar and MOSFET power devices. The basic idea will be to combine in one package the salient features of *both* device technologies. One may, for example, get low switching and low on losses.

Enough data have been gathered to show that power MOSFETs can endure well in the space and military environments requiring radiation-hardened systems. Although sufficient dosage will ultimately affect the parameters of this device adversely, the general indication is that some other system component or part is much more likely to be the weak link in leading to eventual system failure. Moreover, power MOSFETs can be designed with topologies and materials that optimize radiation tolerance with acceptable trade-offs in other characteristics.

Thyristors—Plain and Fancy

The control of many kilowatts of low-frequency (usually 60-Hz) power is likely to remain best accomplished by phase-controlled SCRs. But so-called inverter-type SCRs with frequency capabilities in the 30-kHz to perhaps 50-kHz region will continue to make inroads in heavy-duty power applications. Although such SCRs sacrifice power capability for speed, they can control higher power levels than possible with bipolar transistors and Darlingtons or with power MOSFETs. Of course, this statement can be somewhat negated by push-pull, bridge, and parallel combinations of the lower power-capability devices. Similarly greater power manipulation stems from the use of bipolars and MOSFETs in polyphase systems. The fact remains, and will likely remain, that the mentioned thyristors will retain their power advantage. Accordingly the phase-controlled SCR should see continued service in the control of high horsepower motors and the inverter-type SCR can be expected to make inroads with such equipment as welders, ultrasonic generators, and induction ovens.

As well known, both bipolar transistors and SCRs have been developed for use in tv deflection circuits. Both can be entirely suitable from both cost and operational standpoints. As yet power MOSFETs with 1,500-volt ratings are not a common commodity; however, it is quite feasible that the MOSFET will soon merit consideration for use in deflection circuits. Indeed, where voltage levels allow, power MOS-FETs have already proved themselves in such applications.

An interesting SCR is the asymmetrical thyristor, which can be fabricated and doped to improve turn-on, turn-off times or forward blocking voltage at the expense of reverse blocking voltage. The trade-off scheme had been known, but the exploitation of this fast-acting device had to await a fortuitous observation of a situation common in many control circuits. This was that a *diode* is often connected across SCRs in control circuits. Whether this diode is designated as a free-wheeling, flyback, or protective diode, its manner of connection (anti-parallel) relieves the SCR from having to block a high reverse voltage. Therefore, the semiconductor designer is free to incorporate manufacturing techniques that sacrifice the reverse voltage blocking ability in exchange for some other desirable feature—usually reduced turn-off time—but reduced forward conduction voltage is often an added benefit. This is definitely the way to fly with regard to high frequency and high efficiency performance in inverters and converters operating in a power region awkward or uneconomical for bipolars or MOSFETs.

An even more sophisticated SCR is the reverse conducting thyristor (RCT). The basic philosophy is exactly the same as with the assymmetrical thyristor just described. However, the antiparallel diode is part of the monolithic structure. Not only does this eliminate the need for an external diode, but the inductance of connecting leads is virtually eliminated. This is an important factor at high operating frequencies and high power levels. This device is bound to receive developmental effort to enhance further its electrical and thermal behavior. The integrated thyristor rectifier (ITR) that has been successfully used for horizontal deflection in color tv sets is essentially this device: the RCT. Similarly discrete SCR-diode pairs have also been used for tv deflection circuits. With such a track record we can expect such thyristors to evolve in both power and frequency capability.

With regard to the gate turn-off thyristor, the on-again, off-again involvement of American semiconductor companies with this thyristor has already been alluded to in this chapter. The most recent RCA types were much better devices than most previous devices. Unfortunately a new obstacle interfered with profitable marketing on this novel thyristor. This was unanticipated competition from advanced bipolars—principally Darlingtons—and also from power MOSFETs. Overseas interest in the GTO remains at a high level and domestic development projects are continuing under wraps. It is not unlikely that the next wave of GTO enthusiasm may well force other solid-state devices to close ranks for mutual survival in certain application areas. Motor control is one of these, and regulated power supplies, inverters, and converters are also candidates for a reliable high-frequency GTO with multikilowatt control capability.

One of the most important thyristors is the triac, despite the fact that this device has not appreciably moved up in frequency capability or current or voltage ratings since its early years. Most triacs are used in 60-Hz applications and find much application in light dimmers, fractional HP motor control, and general-purpose control of ac loads such as heater elements, contactors, and solenoids. The 400-Hz types have proved useful in airborne applications, and at least some device and circuit development has resulted in somewhat limited success in the 1.2-kHz region. It hasn't been common to see triacs specified for voltages in excess of 1,200 or for currents beyond about 40 amperes. It is tantalizing that this device, with its desirable features—single-gate actuation and full-wave control of load current—hasn't yet experienced some dramatic technological breakthrough freeing it from its performance constraints. Without indulging too heavily in wishful thinking let's at least hope that designers will include in their device-arsenal 25-kHz and 250-ampere triacs.

Some recently developed thyristors seem destined for greater prominence. This is particularly true with light-fired SCRs, especially now that there have been significant improvements in semiconductor lasers and in fiber optics. Such thyristors obviously provide the best electrical isolation in high-voltage applications. Conversely optical gates are also convenient in triacs, and we are likely to see more devices of this nature assume control functions in mundane applications. In general, much of the improvements to be expected in phase control and inverter-type SCRs will derive from specialized gate structures. So-called amplifying gates, field-effect gates, and gates of various geometries are already optimizing certain SCR parameters, primarily turn-on and turn-off speed. Inasmuch as giant SCRs are already available with 4,000-volt and 5,000-ampere ratings, near-future increases in their power capabilities are likely to be more asymptotic than dramatic.

Thyristors are likely to see more applications using burst modulation because this technique produces relatively little rfi and emi and is easy on such loads as lamp filaments and motor windings. Although the focus here is on circuit technique rather than device technology, it is not too far-fetched to anticipate that manufacturers will find it expedient to optimize certain thyristors specifically for this type of application.

Giant JFETS

Other power-processing solid-state devices can be expected to provide greater versatility for designers. Giant junction field-effect transistors (JFETS and SITs for static-induction transistors) may compete in some areas with MOSFETs. In the recent past much enthusiasm arose over the Class D switching audio amplifier used in some imported stereo equipment. Both power output and general performance were commendable. And the size and weight of this amplifier, together with its heat dissipation, were considerably lower than in conventional Class AB amplifiers. These stereos used large JFETS. Unfortunately the venture ultimately became a marketing fiasco because the JFETS tended to self-destroy from strong driving signals. This does not mean that the problems cannot be resolved; probably the solution involves mutual contribution from improved device technology and better circuitry.

The Schottky Diode

The Schottky diode, although a passive device, has been extremely valuable in inverters and regulators because of its near-instantaneous reverse-recovery characteristic. Additionally the relatively low forward voltage drop in these rectifiers contributes significantly to overall efficiency in 5-volt regulated power supplies, such as those commonly used for computer systems. But a shortcoming of the Schottky diode has been its bad behavior at higher voltages. Its reverse leakage current mounts up rapidly with increased voltage—this has limited application to 40 or 50 volts. As might be expected, laboratories all over the world have been searching for ways to produce a low-leakage 100-volt Schottky. Considerable progress has been made, and probably some diodes of this capability are already in use in the military or in other specialized systems. It is likely that the 100-volt Schottky diode will one day become a common component.

Finale—The Role of ICs and Semiconductor Materials

In the overall picture of solid-state power control it is becoming increasingly apparent that a harmonious working relationship has and will continue to develop between discrete and integrated devices. Thus, we can expect more emphasis on pulse-width modulation ICs, zero-voltage switching ICs, opto-isolators, IC regulators, and hybrid and monolithic power devices with built-in functions such as drive stages, thermal protection, over-voltage and over-current protection, oscillators, voltage references, auxiliary dc supplies, dead-zone circuitry, and various logic functions.

Finally, silicon appears to retain the best *blend* of cost, fabrication qualities, temperature behavior, operational speed, and availability. Germanium still has its adherents and may merit consideration for low-frequency applications and where it is feasible to employ appropriate heat-removal hardware. Gallium arsenide is more elegant than mundane and has been a prominent material for microwave applications. Because of its ability for operation at elevated temperatures, this semiconductor material may surprisingly be found in certain low-frequency power devices of the future.

Index